ECONOMIC INTEGRATION AND GROWTH IN EUROPE

ECONOMIC INTEGRATION AND GROWTH IN EUROPE

A.J. MARQUES MENDES

CROOM HELM
London • Sydney • Wolfeboro, New Hampshire

Croom Helm Ltd, Provident House, Burrell Row,
Beckenham, Kent, BR3 1AT

Croom Helm Australia, 44-50 Waterloo Road,
North Ryde, 2113, New South Wales

British Library Cataloguing in Publication Data

Mendes, A.J. Marques
Economic integration and growth in Europe.
1. Europe — Economic integration
2. European Economic Community Countries
— Economic policy
I. Title
337.1'42 HC241.2

ISBN 0-7099-4664-3

Croom Helm, 27 South Main Street,
Wolfeboro, New Hampshire 03894-2069, USA

Library of Congress Cataloging-in-Publication Data

Mendes, A.J. Marques.
Economic integration and growth in Europe.
Bibliography: p.
Includes index.
1. Europe — Economic integration. 2. European
Economic Community. I. Title.
HC241.M44 1987 337.1'4 86-19937
ISBN 0-7099-4664-3

Printed and bound in Great Britain
by Billing & Sons Limited, Worcester.

CONTENTS

TABLES

To Ana and Miguel
with love

ABBREVIATIONS

CAP - Common Agricultural Policy
CEC - Commission of the European Communities
CEPG - Cambridge Economic Policy Group
CES - Constant Elasticity of Substitution
CIF - Prices including costs, insurance and freight
DG II - CEC's Directorate General II
Dk. - Denmark
EC - European Community
ECU - European Currency Unity
EDF - European Development Fund
EEC - European Economic Community
EFTA - European Free Trade Association
EMS - European Monetary System
Euratom - European Atomic Energy Community
EUROSTAT - Statistical Office of the European Communities
FEOGA - European Agricultural Guidance and Guarantee Fund
FOB - Free on board prices
GATT - General Agreement on Tariffs and Trade
GDP - Gross Domestic Product
GNP - Gross National Product
HMSO - Her Majesty's Stationery Office
INIC - Instituto Nacional de Investigação Cientifica
Irl. - Ireland
MCA - Monetary Compensatory Amounts
NIMEXE - Nomenclature of goods for the external trade statistics of the Community
OECD - Organization for Economic Cooperation and Development
PPS - Purchasing Power Standards
R.H.S. - Right hand side
ROW - Rest of the world
SITC - Standard International Trade Classification
TC - Trade creation

TD - Trade diversion
U.K. - United Kingdom
U.S. - United States
USA - United States of America
VAT - Value Added Tax

PREFACE

This book deals with the problems of measuring the impact of economic integration upon the growth rate of income in the member economies.

It departs from the widespread dissatisfaction with the conventional customs union approach, adding new reasons to explain why this cannot provide a true measurement of the relative gains and losses.

It then presents an alternative, new approach based on the foreign trade multiplier and the hypothesis that in the long run growth is constrained by the balance of payments. The framework allows a direct estimation of the integration effects upon GDP and provides a splitting of the various mechanisms through which the effects operate, namely: changes in the terms of trade; the propensity to import; the export growth; factor mobility and the operation of common policies.

The new approach requires the estimation of induced effects on external trade flows; the weighted share technique has been selected to make these estimates. However, we extended it considerably in order to overcome some of its limitations and provide a more realistic "anti-monde" to compare with the actual values.

We could then proceed with an evaluation of the European experience during the period 1958-1981. Detailed estimates are presented for both the EEC and EFTA for the two basic periods, ie, before and after 1973, and disaggregated at the country level. This facilitates a comparison of their relative growth performance, provides an answer to some of the theoretical questions raised in other studies and present a wide ranging set of conclusions and policy issues. As usual, many questions remain unanswered but several new directions for future research were identified.

A global conclusion is that up to 1981 the countries involved in the European integration process had a Gross Domestic Product gain equal to at least 1.5% of their GDP in 1981.

The author wishes to acknowledge the many friends that made this book possible and the financial support from the EEC Commission and the EFTA Secretariat. In particular, my deepest gratitude goes to Prof. A. P. Thirlwall, who, as the leading advocate of the balance of payments constrained growth hypothesis, inspired my interest in the subject and throughout my work was enormously patient to revise and comment the various manuscripts submitted to him. Last, but not least, I would like to thank Paula for her help not just with the bibliography but also for taking over all the family responsibilities during my prolonged absences from home. She and the children (Ana and Miguel) were the main victims of these absences whose consequences we all have to bear in the future, and I can only trust in their great love and care and hope to be forgiven.

As usual any remaining errors and shortcomings are my own fault.

Canterbury, February 1986

A.J. Marques Mendes

A poem by Camoes

Ves aqui a grande máquina do Mundo

Etérea e elemental, que fabricada

Assi foi do Saber, alto e profundo,

Que é sem principio e meta limitada.

Quem cerca em derredor este rotundo

Globo e sua superficie tao limada,

E Deus; mas o que é Deus, ninguém o entende,

Que a tanto o engenho humano nao se estende.

In: Os Lusiadas, Canto X.80 - first published in 1572.

An English translation by Sir Richard Fanshawe - published in 1655

The WORLD'S great *Fabrick* thou doest *heer* descry
Heav'nly and *Elementall:* for just so
'Twas made, by that *All-wisdome,* that *All-eye,*
Which no *beginning* knew, no *end* shall know:
Which *interweaved* in each *part* doth lye,
And round the fair *Work* like a *Border* goe:
'Tis GOD. But what God is, poses *Man's* wit,
Nor can *short Line* fathome the INFINIT.

Chapter One

INTRODUCTION

The original motivation for this book was to gain some insight into the likely effects of further enlargements of the EEC. The initial presumption was that it would be much easier to study membership of an existing union, with a large "acquis communautaire" built during a period of more than 25 years, than the question of forming a new one. After all, the alternative of remaining outside had already been experienced.

However, a brief survey of the literature suggested otherwise. Broadly speaking the studies can be divided into two main groups. One that we might call "the historical perspective", focuses on the comparison of the relative achievements of individual member countries and the implementation of the various common policies and institutions. A second, and much larger group, relies on the simplification that economic integration can be reduced to the formation of a customs union and, therefore, is exclusively concerned with the corresponding theory and the measurement of its trade effects. Needless to say, neither of these two groups of literature is conclusive when it comes to assessing the individual advantages of integration.

The concept of integration itself, although widely used, is quite new (cf. Machlup 1977), and has still not got a well-defined contour. The two basic ideas associated with the concept are the need to abolish discrimination within a given spatial[1] unit and the necessity of promoting some sort of policy coordination on issues considered as being of mutual interest.

Bearing in mind the complexity of these concepts, Balassa (1961) was led to propose a very broad definition of: "integration as both a process and a state of affairs". Writing twenty-three years later, Robson (1984) emphasises that: "international economic integration is a means, not an end".

However, it has also been widely accepted that the theory of economic integration, although being a part of international economics, extends beyond customs union theory in several important respects, namely:

(a) by taking into account international factor movements and elements of location theory; (b) by setting up some degree of policy coordination in areas other than trade policy, e.g. in the monetary and fiscal field; (c) by using various indicators to measure the effect of "unifying" separate economies, and (d) by assessing their contribution to the achievement of pre-established political objectives.

On the other hand, there is also widespread acceptance of the various forms that the integration processes may take. They are:

Free Trade Areas, where member countries agree on reducing intra-area restrictions to trade, while still keeping individual protection schemes against third countries;

Customs Unions, where apart from liberalization of intra-union trade, the members also establish a common customs tariff against third countries;

Common Markets, where the reduction of trade restrictions is also extended to factor movements;

Economic Unions, where apart from reducing restrictions on trade and factor mobility there is also some harmonization of national economic policies to cut remaining sources of discrimination; and

Unification, which presupposes a supra-national authority with power to pursue unified policies in various areas, including monetary; fiscal and social policies.

It is not necessarily the case that any of these various forms represents a phase towards a unification process. Each one of them can constitute the ultimate goal in itself. Furthermore, for higher stages of integration there is no clear boundary to define when one form ends and the other begins[2], e.g. the boundary between a common market and an economic union. The process of European integration clearly illustrates these two points because on the one hand we have the European Free Trade Association (EFTA), whose ultimate goal was the establishment of a free trade area, and on the other hand we have the European Economic Community (EEC) which is invariably designated as a common market or economic union.

In the EEC context a new argument has recently emerged[3] concerning the question of the convergence (divergence) required (tolerated) to implement the economic union. The debate arose originally in regard to the monetary union schemes and has consequently centered on the questions of inflation, current account balances and stabilization policies; but there is certainly underlying and long-lasting concern about the effects of integration upon the per capita income of the member countries [see: Marjolin (1975); MacDougall (1977); Hallet (1982); Steinherr (1984)].

This last concern is, in fact at the heart of the literature which utilises the "historical perspective" to evaluate the integration process with a view to answering two basic questions: (1) has Europe become more or less equal?, and (2) what has been the role of integration in this process? A preliminary question should also be considered: has Europe grown faster than other similar economies as a result of integration[4]?

However, the existing literature is far from conclusive on any of the above questions, regardless of the regional aggregates used. The studies can be split up into those suggesting that inequality might have increased [e.g. Seers (1979); Kiljunen (1980); Secchi (1982) and Denton (1982)]; those suggesting a reduction of inequality [e.g. Whitbread (1981); Biehl (1978), and Hallet (1982)], and those that are inconclusive [e.g. Molle et al (1980) and Keeble et al (1982)]. The reasons for such disparate predictions derive from both the inherent difficulty of the questions and from deficiencies and/or inadequacies of the methodologies used. In a non-exhaustive enumeration of these we can certainly include the absence of a precise definition of what is meant by divergence; the difficult choice of an appropriate inequality measure; the use of only a few point estimates to deduce trends; the non-comparability of data bases, and the absence of a control group of countries by which comparisons might be made.

We can use table 1.1 below to illustrate some of these questions.

Table 1.1 - Comparison of income levels and growth rates

Countries	GDP per Capita in ECU's at 1980 prices and *PPS*[1]					GDP growth rates annual averages			
	1960	1965-72	1973-76	1977-81	1983	1960–65[2]	1965-72	1973-76	1977-81
Spain	..	4095	5426	5698	5804	..	6.33	3.27	0.99
Portugal	1384	2329	3164	3555	..	6.46	6.62	1.28	3.76
Greece	1575	2831	3858	4374	4389	8.06	7.46	2.95	2.95
Ireland	..	4013[3]	4425	5113	5365	..	4.68[3]	3.19	4.06
Denmark	..	7026[4]	7992	8684	9186	..	3.96[4]	1.70	0.99
U.K.	5055	6119	7038	7482	7803	3.19	2.47	0.68	0.62
Netherl.	4477	6069	7394	8325	8156	4.87	5.02	2.65	1.29
Italy	3362	5020	6133	6789	6937	5.20	5.14	2.12	2.92
Germany	4865	6534	7768	8817	9128	5.01	4.09	1.49	2.33
France	4220	6120	7699	8614	8975	5.76	5.45	2.78	2.21
Belgium	4167	5915	7531	8205	8518	5.00	4.75	2.65	1.82

Source of data: EUROSTAT - National Accounts, Aggregates 1960-83 and 1960-81

Notes: 1) PPS - Purchasing Power Standards
2) Based only on the two terminal years
3) Period 1970-72
4) Period 1966-72

The first point to note is that there are very large differences between the growth achieved by individual countries, which for a given period might be in the order of one to six. Secondly, there is also a wide disparity in per capita incomes in the order of one to four. However, it is not obvious whether the various values are indicative of a trend towards divergence or convergence.

To begin with the concept of convergence/divergence in itself must be clearly stated because inter-country comparisons should be made in terms of both growth rates and levels as well as in relative and absolute values. Furthermore, convergence/divergence must be looked at not just as a simple two-point measure of disparities but as a movement towards greater equality. This means that we should envisage a process whereby we depart from a situation of inequality by reducing the relative differences until we reach a position where absolute differences start to be reduced. Only then can we really talk about a convergent process. However, we can as well define weak (relative) convergence as a situation where a deceleration in the growth of the absolute income gap begins.

An evaluation of the European performance on these lines can be appreciated through table 1.2, below. It can be seen that although there are significant reductions in the relative differences as far as

Table 1.2 - Comparative income performance relative to the EEC8 average

Countries	1953-1958		1959-1964		1965-1972		1973-1976		1977-1981	
	1	2	1	2	1	2	1	2	1	2
Spain	234	2.61	233	3.79	202	2.22	185	1.55	194	3.32
Portugal	468	3.26	437	4.06	375	2.63	336	2.73	329	0.99
Greece	356	3.23	334	3.60	269	1.63	234	2.13	229	1.93
Ireland	193	6.76	206	4.44	210	3.97	211	2.99	207	1.00
Denmark	69	-1.60	69	6.59	68	3.83	71	0.58	73	-1.36
U.K.	100	61.71	109	22.03	122	14.04	130	3.79	135	5.46
Netherl.	82	-1.89	87	3.46	85	7.08	86	-3.82	87	0.96
Italy	172	2.63	160	3.41	154	3.09	153	2.14	154	1.09
Germany	80	11.61	77	5.67	78	3.78	78	2.50	76	4.34
France	93	7.34	93	7.81	88	12.42	84	5.74	83	2.61
Belgium	86	-7.57	92	6.10	90	8.07	84	12.91	86	-1.21

Source of data: OECD - Main Economic Indicators, several issues.

Notes:

1) Ratio EEC8/other countries of per capita income (annual average) x 100

2) Percentage change (annual average) of the income gap between each country's per capita income and the EEC average income.

the absolute income gaps are concerned, only a few countries show a reduction. Moreover, from the mid 1960's onwards there was a slowing down in the reduction of relative differences.

Another important point to note is that not all countries seem to contribute to this convergence. We have a very mixed situation where countries with incomes both below and above average (U.K., Ireland, France and Germany) deviate from the trend, while we also have some convergence to it (e.g. the Mediterranean countries, Denmark and Netherlands). Therefore, we cannot be conclusive concerning the overall trend unless we have a global measure of the income distribution[5].

But, as earlier studies have shown (e.g. Molle et al 1980), the choice of inequality measure is not a simple issue. We can only rely on comparing two inequality measurements if we are sure that the corresponding Lorenz curves do not cross. Otherwise we cannot be certain whether the observed constancy or change of the inequality measures was uniform or due to offsetting changes in the specific segments of the income distribution. With regard to this, Atkinson's (1970) measure seems preferable since it specifically introduces the weight (ϵ) that society puts on inequality. In fact, the other measures have also embodied some form of a social welfare function, which in the case of the Gini coefficient gives more weight to transfers affecting middle income classes and in the coefficient of variation puts equal weight on transfers at different income levels. However, with Atkinson's measure when we give ϵ a value of one we know that we are regarding it as being "fair" to take (for example) $1.00 from the rich and give $0.50 to the poor (or $0.25 in the case of $\epsilon = 2$). So, by increasing ϵ we are revealing our aversion to inequality and we can always use it as a measure for intertemporal comparisons.

Illustrative examples can be taken from table 1.3 below, which shows the inequality measures and rankings using the Atkinson coefficient, the Gini coefficient and the weighted coefficient of variation. Indeed, from Atkinson's measures for 1957 we can see that in this pre-integration year inequality was higher in the lower income segment, as is shown by the higher ranking obtained when a greater weight ($\epsilon = 2$) is given to equality in this segment of the distribution. Also, using column 4 we can illustrate how misleading might be the use of only two extreme years to draw conclusions in relation to trends in convergence/divergence. In fact, taking 1956 as the base, we would conclude that income disparities had remained constant, slightly increased (1%) or decreased (3%) depending on whether we use 1977, 1978 or 1976 as the comparison year. We must therefore conclude, that a trend can only be taken from a series such as is shown here, which, in spite of the problem of "crossed" distributions, clearly confirms the indication found in some of the studies surveyed that since the mid 1960's there has been a divergent growth of real income per capita between the EEC member countries.

However, despite the fact that we can state that there is no doubt that these economies are diverging in terms of income, this is not tantamount to saying that integration has played a role in this divergent process. A preliminary, and very simplistic, way of checking this question is the matching of their performance against that of a

Table 1.3 - Inequality measures between EEC8 countries and their ranking.

Years	Value	Rank	Value	Rank	Value	Rank	Value	Rank	Value	Rank
1953	.1266	29	.0371	19	.0580	16	.0801	14	.2410	27
1954	.1312	28	.0391	13	.0613	9	.0847	8	.2470	22
1955	.1404	20	.0407	8	.0635	7	.0874	5	.2550	17
1956	.1428	17	.0409	7	.0636	6	.0874	5	.2580	13
1957	.1373	22	.0385	14	.0600	13	.0826	10	.2500	19
1958	.1368	23	.0362	20	.0561	20	.0768	19	.2470	22
1959	.1368	23	.0351	25	.0541	24	.0739	22	.2460	24
1960	.1435	15	.0377	16	.0581	15	.0791	16	.2570	15
1961	.1407	19	.0352	24	.0539	25	.0731	24	.2500	19
1962	.1394	21	.0337	26	.0515	26	.0697	26	.2460	24
1963	.1352	26	.0320	28	.0490	28	.0663	27	.2400	28
1964	.1416	18	.0356	23	.0545	22	.0738	23	.2500	19
1965	.1464	12	.0376	18	.0575	18	.0778	17	.2580	13
1966	.1430	16	.0361	21	.0551	21	.0745	21	.2540	18
1967	.1332	27	.0314	29	.0479	29	.0648	29	.2370	29
1968	.1367	25	.0323	27	.0491	27	.0660	28	.2420	26
1969	.1447	13	.0360	22	.0544	23	.0728	25	.2560	16
1970	.1473	10	.0377	16	.0570	19	.0763	20	.2630	11
1971	.1486	8	.0392	12	.0594	14	.0795	15	.2660	10
1972	.1493	7	.0404	9	.0613	9	.0820	11	.2690	7
1973	.1480	9	.0398	11	.0604	12	.0809	13	.2680	8
1974	.1442	14	.0383	15	.0580	16	.0776	18	.2630	11
1975	.1466	11	.0404	9	.0612	11	.0820	11	.2680	8
1976	.1505	6	.0417	6	.0631	8	.0844	9	.2740	6
1977	.1529	5	.0432	5	.0654	5	.0874	5	.2780	5
1978	.1539	4	.0436	4	.0660	4	.0882	3	.2800	4
1979	.1561	2	.0443	2	.0669	2	.0892	2	.2840	3
1980	.1553	3	.0439	3	.0662	3	.0882	3	.2850	2
1981	.1567	1	.0448	1	.0675	1	.0898	1	.2880	1
		1		2		3		4		5

Source of data: OECD - Main economic indicators - several issues.
Rankings according to the value of: 1 - Gini coef.; 2 - Atkinson coef. with $\epsilon=1$; 3 - with $\epsilon = 1.5$; 4 - with $\epsilon = 2$; 5 - Weighted coef. of variation
Note: - EEC8 includes all the countries belonging to EEC after the first enlargement, except Luxembourg which was excluded.

similar[6] control group of non-member countries.

Obviously, the use of a control group is not without difficulties (after all there are not two identical economies), but it has some validity in that here we have been able to deal with a long period of time and various groups. A selection[7] was made in order to constitute two groups of countries that, in terms of population, GDP and income per capita, were similar to the EEC with five and eight members (Luxembourg was excluded because of its size). However, a simple comparison of the time patterns is bound to be concealed by differences in variables that underlie the evolution of the inequality measures. Since the literature on economic development [e.g. Kuznets (1963) and Chenery et al (1979)] has shown the existence of a link between the level of development and the income distribution we also regressed our inequality measures against the per capita income for both the EEC and the control groups. Moreover, since a relation has already been found from the income distribution across countries that points to an inverted U form, we adopted a quadratic specification[8] to make our estimate.

The results for the Atkinson measure with a weight of two are illustrated graphically in figures 1.1 and 1.2 below. Five points stand out: (1) - Both the EEC and non-EEC countries present the same pattern in terms of convergence/divergence; (2) - the velocity of change in inequality is greater in the non-EEC countries; (3) the difference in inequality between the EEC set and the control group has been reduced for both the ante and post-integration periods, with a deterioration of the EEC relative position (note that the EEC5 became even more unequal); (4) in both periods the control groups experienced a faster growth of income; and (5) the trend for divergence (increasing inequality) appeared earlier (in time) for the EEC countries. It began around 1967 for both the EEC5 and EEC8, while for the control5 and control11 groups it began in 1971 and 1978, respectively.

We can now advance some explanations for these observations. A basic and straight-forward answer is that the EEC itself has played a significant role in explaining the greater divergence between the EEC countries. However, it could also be argued that this divergence was entirely due to the EEC commencing from a position of lower inequality and higher income combined with lower growth rates. Moreover, there is the fact that both the EEC5 and EEC8 began diverging at the same time while the later was only formed in 1973. It may also be said that due to some harmonization induced by integration this group is less sensitive to cyclical variations which (arguably) were the main source of divergence. The counter-argument is that divergence for the control groups started on the eve of a slowdown (oil crisis), while in the EEC it started in a normal period (which is suggestive that integration was a source of divergence). However, further evidence is clearly needed. One test, which we applied, is the introduction of a dummy variable (taking the value 0 and 1 for - ante and post - integration periods respectively) in order to assess if there was any shift of the function. This procedure was also adopted for the control groups in order to check whether the (possible) shift could be ascribed to integration. Of these regressions, only in the case of the EEC5 there was a (positive) significant parameter.

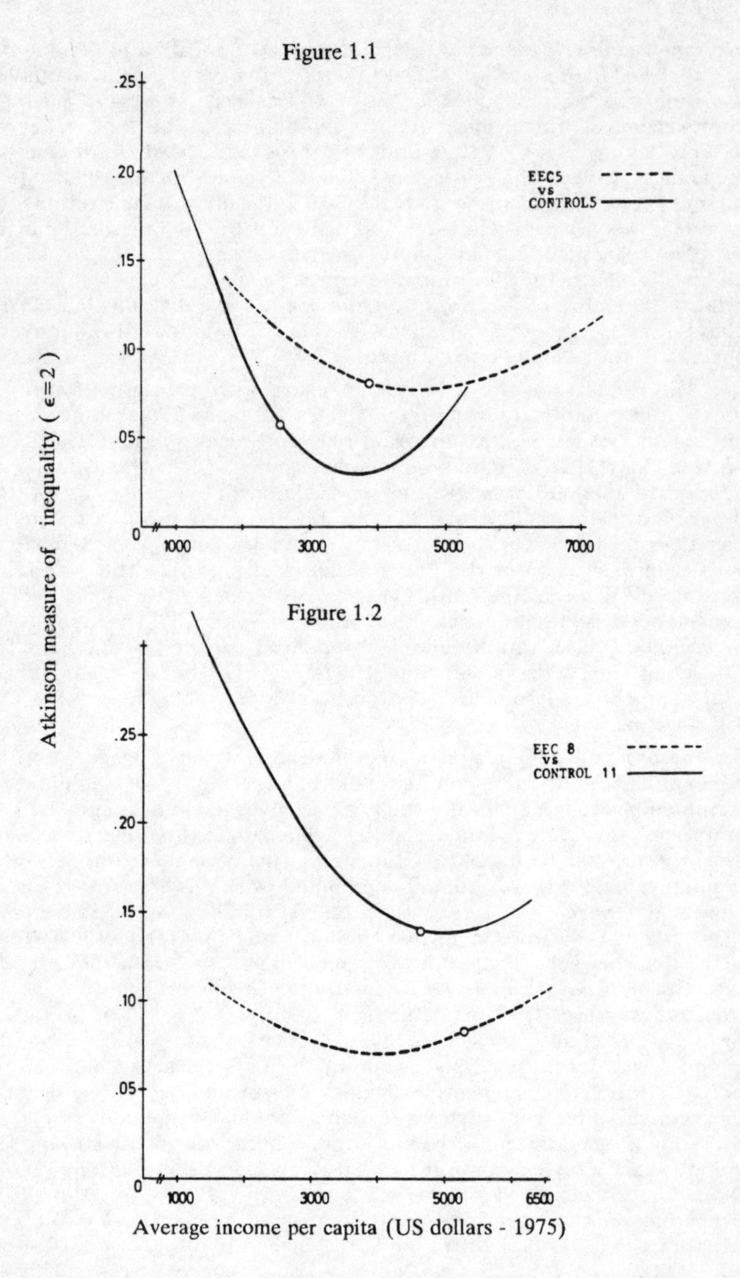

Notes: Dark lines show the period of estimation. Dots show years of effective establishment (1964) and enlargement (1976) of EEC.

But, if we also add the fact that for the post-enlargement period the dummies had a positive sign for the EEC and a negative sign for the control group, then there is also some (but weak) ground to suggest that integration might have induced some divergence.

However, this evidence is clearly insufficient to draw any definitive conclusion on what might have been the role of integration in the trend towards divergence. All we can conclude from the literature and our own research within this "historical perspective" framework is that: (1) a divergent pattern can be identified for the EEC, both before and after enlargement, which starts in the mid 1960's; (2) the EEC performance in terms of a reduction of inequalities is much worse than that of the comparison groups, which raises serious questions regarding the role of integration and the efficiency of EEC policies in promoting equality; (3) both the country by country and the global analyses are consistent with the cumulative causation predictions, although they cannot provide conclusive evidence for this hypothesis. Whatever, the view on these questions, what is unquestionable is that there is no reason at all for new members (Iberian or others) to prejudge that their national economies will improve either absolutely or relatively as a "natural" consequence of membership. What is needed is a methodology that can embrace the various mechanisms through which integration might affect the member countries' economies.

Customs Union Theory, which underlies the bulk of literature on international economic integration, can only be looked to provide a valid evaluation of the effect of integration if we are also prepared to accept the tenet of this theory that integration can be simplified in the formation of such a Union. A further problem in this theory arises from the fact that an odd paradox exists: while attempting to assess gains made from forming a customs union, the theory sets out to make an empirical assessment while at the same time, the measurements to be used are not operational. Because of this there is an operational deadlock. Therefore, in this book the main task that we set ourselves is to attempt the development of a new and alternative way of measuring the effects of integration. In doing so we will adopt the view that "a primary economic incentive for states to enter into arrangements for integration is the prospect of economic gain, in the shape of an increase in the level or rate of growth of output" (Robson 1984), and we will try to measure the effects of integration on the Gross Domestic Product of each member country.

The book is basically divided in two parts. The first part (chapters II,III,IV and part of V) deals with theoretical developments required by the balance of payments framework to measure integration effects, while the second part (chapters V,VI,VII and part of III) contains the empirical work needed to use the suggested methodology to evaluate the European experience of economic integration. We begin in Chapter II therefore with a brief presentation of the "state of art" of customs union theory, followed by an examination of its major drawbacks both as a branch of pure positive economics, relying extensively on general equilibrium assumptions, and its exclusive reliance on trade creation/diversion effects which we show to be hardly (if at all) measurable empirically. The possibility of dynamic effects of integration occurring is also examined. Having concluded that it is simpler and more correct to measure the effects of integration on total trade and on the balance of payments we proceed in Chapter III with the presentation of an alternative framework by which to measure integration effects. The starting point is the foreign trade multiplier which was derived long ago by Harrod (1933) and which has been recently revived and made dynamic by Thirlwall (1979,1982) in his balance of payments constrained growth model. This framework is briefly outlined and illustrated with estimates made for the EEC and EFTA countries. Starting with the overall balance of payments identity we shall then derive the framework required to estimate the integration effects on the growth rate and split them up into their various components, namely: changes in the terms of trade; the propensity to import; export growth, and factor mobility. The model requires the estimation of the trade effects induced by integration. In Chapter IV, we give a brief review of the various techniques that have been used for doing this we decided to opt for the weighted share methodology (e.g. Verdoorn and Schwartz 1972) and we give a clarified presentation of this. Furthermore, we extended the methodology in order to provide a more realistic "anti-monde", including the introduction of the home market and the possibility that if, in the absence of integration, the various countries had kept constant their relative growth rates of import and export then the trade effects would have remained unchanged (a result proved in Appendix to chapter IV).

In the second part we proceed (Chapter V) with the estimates of the trade effects for the EEC covering the period before and after the 1973 enlargement. Apart from the estimation of trade effects this chapter also includes a discussion of the likely impact of the Common Agricultural Policy on the balance of payments, and it is argued that the conventional measurement of the increase in the price of foodstuffs and the budgetary net contributions should be added to the effects on trade of food commodities. The total net budgetary position of each country and the effects resulting from factor mobility are also presented. In particular, the hypothesis of tariff discrimination effects on direct foreign investment is considered. These various estimates are then used in Chapter VI to measure the impact on the growth rate using the balance of payments framework. Finally, in Chapter VII similar estimates are presented for EFTA, whose performance is compared with that of the EEC. The overall impact on European growth and the effects on inequality are also measured, which then allow a clarification of the doubts raised in the introduction concerning the effect of integration on divergence shown between the European economies.

The various conclusions and policy issues obtained are referred to throughout the text. The same applies to insufficiencies and new directions for research identified during our work. However, a brief overall summary of the main conclusions and new directions for future research is given at the end (Chapter VIII).

1. One should distinguish between spatial levels, namely: regional; international and global. However, we will only retain as relevant for the present study the international or inter-country level.
2. A very up-to-date example of this is given by the editors [Bieber et al (1985)] of "An Ever Closer Union", which states that "the term European Union is delightfully ambiguous" and that "Empirically, one might as well abandon any hope of arriving at a common meaning of the term".
3. In fact, there is hardly a new report from the EEC that does not include at least one paragraph on the question of convergence/divergence in the Community.
4. Looking at developments which occurred during the twenty-three years between 1960 and 1983, the plain answer is that the average per capita income in terms of purchasing power standards has increased by 88.4% in the EEC10 grouping against an increase of 55.2% in the United States and 163.5% in Japan (between 1965-83). However, this is not tantamount to saying that integration played a major part in this outcome.
5. Furthermore, an attempt made to search for some common factor to explain each country's achievement, using the country's size (population), the level of development and the trade position before integration did not prove conclusive either.
6. Surprisingly, none of the literature reviewed, which produces deductions/opinions on the role of the EEC, has ever compared its performance to other countries.
7. In the case of EEC5 it was even possible to establish a control group in which the countries, to some extent, matched each other (say Netherlands-Austria; Italy-Spain; Germany-Japan; France-U.K. and Belgium-Denmark). For the EEC8 a larger group had to be taken including all non-EEC country members of the OECD, with the exclusion of the United States, Turkey, Australia and New Zealand. In the period 1960-69 the control groups presented the following average values (the EEC figures are in brackets): Control5 with a population of 195(180) million people had a GDP of 525(720) billion of 1975 U.S. dollars and an average GDP per capita of 2686(3998) U.S. dollars; control11 (and EEC8) with a population of 181(242) million people had a GDP of 563(934) billion of 1975 U.S. dollars and an average GDP per capita of 3100(3862) U.S. dollars.
8. It must be noted that Kuznets' (1963) pioneer work and further studies were effected by international comparison of inequality within countries. However, other specifications were tried but with less satisfactory results, namely, a quadratic plus a time trend, a hyperbola, logarithmic and a polynomial of degree three.

Chapter Two

CUSTOMS UNION THEORY AND ECONOMIC INTEGRATION

2.1 - Customs Union Theory - A brief "state of the art"

The pioneer work in the field of customs union theory was Jacob Viner's book on "The Customs Union Issue" (1950). Since then fundamental contributions have been made by Meade (1955), Lipsey (1960), Johnson (1965) and Corden (1972). There is now a vast amount of literature presenting the partial and general equilibrium versions of the theory. Here we will follow Jones's (1981) model of two commodities and three countries because it allows the presentation of the general equilibrium version of the theory in a diagramatic supply - demand framework. Let us then start with a world economy made up of three countries, that is, a "home country" plus a "partner country" willing to form a union and a "rest of the world country", and assume[1] that the many commodities being produced can be reduced to two composite goods [1]. They could represent agriculture and industry; importables and exportables or, alternatively, that there is a single homogeneous product and a second good which is an aggregate of "all other goods". This last case can be interpreted as the partial equilibrium approach if the single sector is conceived as being so small that any changes affecting the sector will not have any effect on prices elsewhere in the economy.

Similarly we can aggregate the production factors so that the economy can be represented by an Edgeworth-Bowley box diagram, as shown in Figure 2.1 below. This clearly depicts one of the basic assumptions of customs union theory centered on the immobility of factor inputs at the international level [2], which substantially reduces its usefulness in the EEC case where the free mobility of factors is one of the basic objectives of the common market and where the tariff-discrimination hypothesis[2] seems to be confirmed (see chapter V). This is further coupled with the assumption of a given and perfectly known technology [3] which shows the static nature of traditional customs union theory.

From the contract curve $(O_x q O_y)$ in the box diagram we can now derive the more conventional curve of the possibilities of production, because the perfect competition assumption [4] ensures that

Fig. 2.1

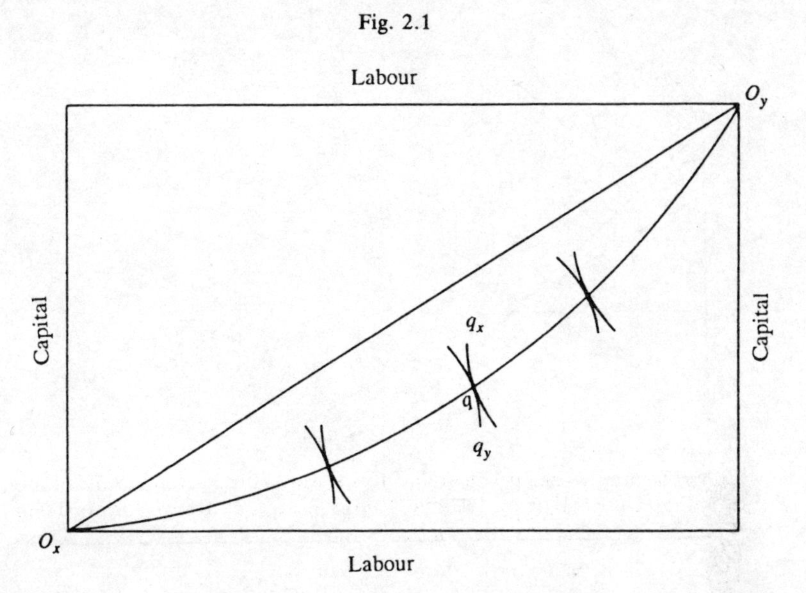

production will always occur along the contract curve.

In figure 2.2 below we depict the case where the production possibility frontier is concave to the origin, that is, the assumption of constant returns to scale[3] [5]. Now, if we use one of the commodities as the numeraire, say importables (q_x) expressed in terms of exportables (q_y), then the relative price will be $p = \frac{q_y}{q_x}$ and we can redraw figure 2.2 as a conventional supply and demand diagram such as in Figure 2.3 below where S_x is the inverse of the marginal cost curve of the importable good and corresponds to the slope of the production possibility frontier. In fact, given the perfect competition assumption the curve can be taken as the domestic supply function for commodity x. In turn, D_x can be interpreted as the equivalent demand curve for importables and is drawn by taking the inverse of the curve linking the tangency points in the commodity space. However this interpretation is more complex. It requires two basic assumptions, namely that consumers are utility maximizers with given preferences [6] and that their consumption plans derive exclusively from income generated from domestic production[4] [7]. This last assumption, coupled with the full-employment assumption depicted in the Edgeworth-Bowley box diagram, must also imply that the balance of payments adjusts automatically without significant costs[5] [8].

Given these assumptions, and calling the economy of figure 2.3 the "home country", let us now consider the effects of establishing a customs union, which implies the elimination of tariffs on imports from the "partner country" and the setting of a Common Customs

Fig. 2.2

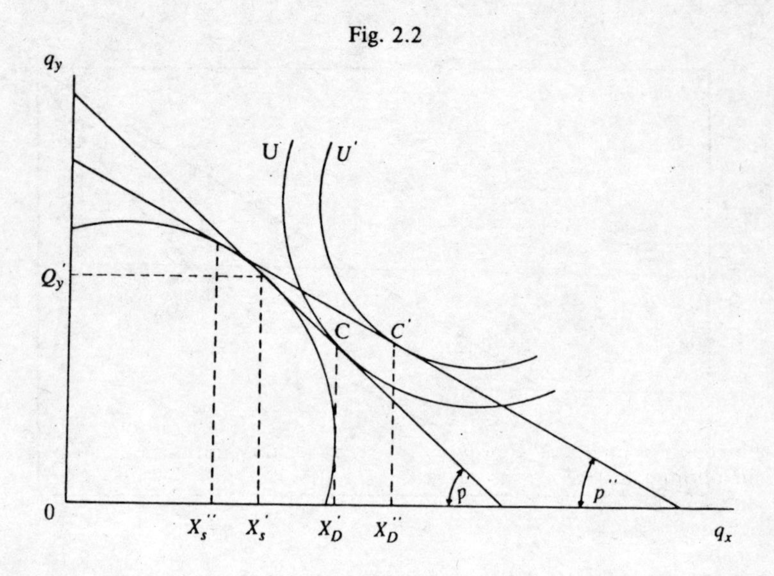
q_y
U
U'
Q_y'
C
C'
p'
p''
0
X_s''
X_s'
X_D'
X_D''
q_x

Fig. 2.3

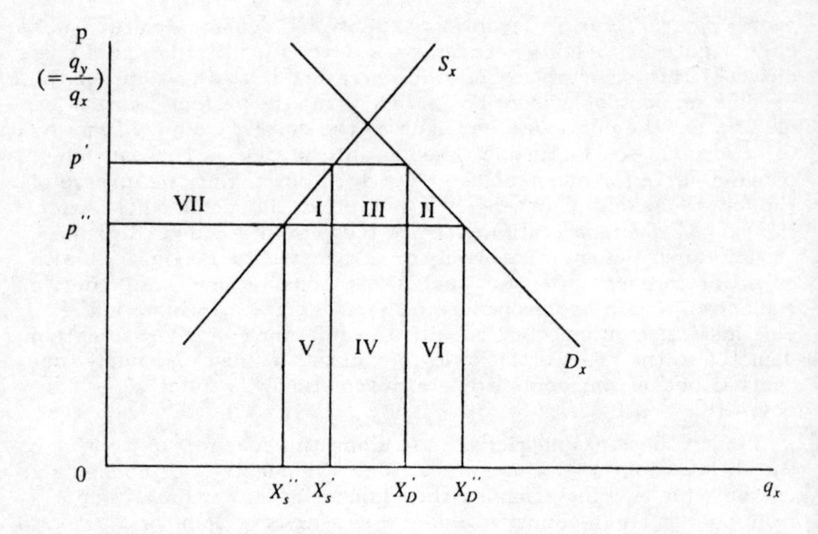
p
$(=\frac{q_y}{q_x})$
S_x
p'
p''
VII
I
III
II
V
IV
VI
D_x
0
X_s''
X_s'
X_D'
X_D''
q_x

Tariff against the "rest of the world" causing a change in the terms of trade from p' to p'' in figure 2.3 above.

First note that total output ($X_s'+Q_y'$ in figure 2.2) at the pre-union terms of trade (p') in figure 2.3 is equal to $2p'X_s'$. Now a change in relative prices due to tariff reduction makes the importable commodity less valuable when measured in terms of exportables. That is, there will be a loss (VII + I) in the case where producers do not change the level of production and VII only if there is production reallocation [VII = (VII + I + V) - (I + V)]. The difference (I) can therefore be seen as a source of gain. This is what Viner (1950) calls trade creation and corresponds to the replacement of inefficiently produced import substitutes for cheaper imported goods. There is also the possibility that the new terms of trade are such that there is not only a substitution of domestic production for partner imports but also a substitution of cheaper imports from the "rest of the world", which Viner classified as trade diversion. The final outcome in terms of welfare will then be the difference between these two effects and the formation of a customs union will only be beneficial if the result is positive. However, as Meade (1955) pointed out long ago, the change in the terms of trade (p) will also have an effect on consumption. Looking back at figure 2.3 we can see that there is an increase of demand for importables ($X_D'X_D''$). The total consumers' surplus resulting from a change of the terms of trade from p' to p'' will be given by areas VII + I + III + II, from which the loss of producers' surplus (VII) must be deducted. It should be noted that the consumption effect (II) is the result of both substitution and income effects, but its translation into welfare gains requires some qualifications. Scitovsky (1958), the first proponent of the use of welfare triangles, pointed out that this requires the assumption of constant returns to scale and a constant exchange rate [9] which are highly questionable. Further, it requires that the given preferences are represented by convex and non-intersecting indifference curves such as in fig 2.2, and the possibility of translating the ordinal concept of welfare implied in these curves into quantities of a commodity [10], such as in figure 2.3, has only been proved (Burns 1973) for the case of one commodity. In fact, the extension of the model to more than two commodities is sensitive to the taxonomy of the customs union effects which will be referred to in the next section.

Meanwhile, let us proceed by considering the third effect (measured by area III) which corresponds to additional income available to the home economy as a result of the lower cost of the original volume of imports. This has been termed the "terms of trade effect" (Jones 1981), but it does not cover all the effects usually referred in the literature under this heading, which can also derive from the exploitation of a stronger bargaining position which keeps the exchange rate from falling and obviates the need for any balance of payments adjustment.

At this point it is important to note that the effects of membership to a Customs Union cannot be measured by considering exclusively the point of view of a single country. In fact, we need to separate the terms of trade effects originating in trade with the "partner" from those of the "rest of the world". However, conventional customs union theory has bypassed this problem by assuming a "small" home country, which cannot influence the terms of trade with any of the other two economies. This case can be illustrated through Fig. 2.4 below where the curves $\overline{p_w^t d^u}$ and $\overline{p_u d^u}$ are, respectively, the (tariff-excluded) terms of trade offered by the "rest of the world" and the "partner", so that the total supply curve is $\overline{ss^t d^t}$.

Fig. 2.4

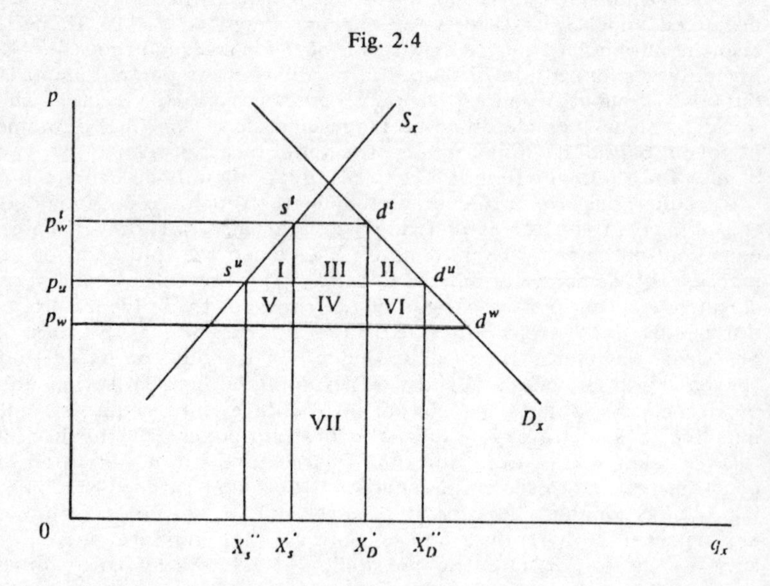

Traditionally, the effects of a change in the terms of trade from p_w^t to p_u are measured by deducting the tariff revenue loss (III + IV) from the economic surplus (I + III + II) obtaining a net effect of (I + II - IV), that is, trade creation (I + II) minus trade diversion (IV).

An alternative way of presenting this result can be done by redrawing this figure as Fig. 2.5 below, where the $\overline{mm^u}$ curve represents the amount of demand of imports at the various terms of trade for the home country. So, at p_w^t imports will be $\overline{ox_m^t}$ (that is, $\overline{ox_m^t} = \overline{s^t d^t}$ in Fig. 2.4), and similarly for the other terms of trade. Now, given the "small-country" assumption the supply elasticities of both the "partner" and the "rest of the world" are infinitely elastic so that we draw the corresponding supply curves (tariff-inclusive) as $\overline{p_u^t p^u}$ and $\overline{p_w^t m^t}$. The numbered areas under the demand curve for imports are easily identified as the counterparts of those presented in supply-demand diagram. The measurement of the trade creation and trade diversion effects is done in a similar way. However, if we reject

Fig. 2.5

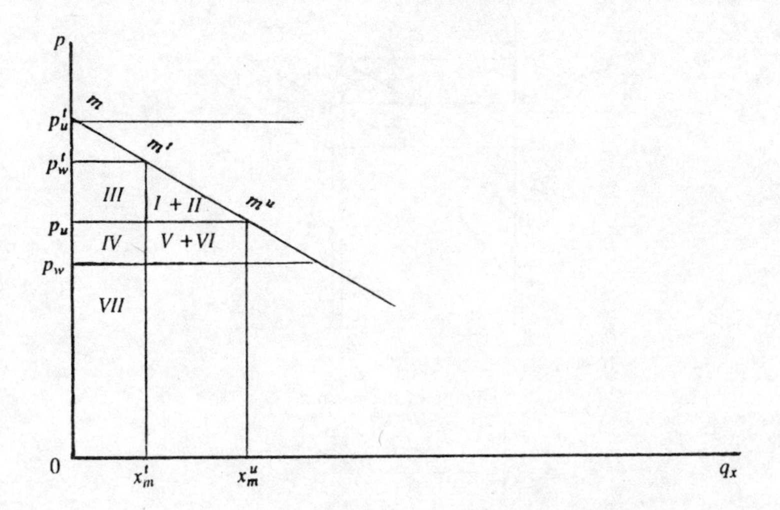

the assumption of infinite elasticity we have to represent the total supply by drawing upward sloping curves, and we need to consider the effects on the three economies simultaneously.

This brings us to a widely encountered dichotomy in customs union theory based on a distinction[6] between the large and the small union cases. Because we are mostly interested in the EEC customs union and the EEC-EFTA free trade area, we shall proceed here by considering only the large union case[7]. Its main feature is that we have to abandon the single [8] country focus and concentrate on examining the combined effects of membership on both the "home country" and the "partner country". We shall start by redrawing figure 2.3 above as figure 2.6 below. This is done by assuming zero transport costs [11] and no other trade barriers. So, we can depict the import demand function (m_X) below corresponding to the amounts of imports shown above (that is, proportionally, $\overline{p^t p^t} = \overline{X_s' X_D'}$ and $\overline{p^u p^u} = \overline{X_s'' X_D''}$) and the export supply function (E_x) corresponding to the sum of the supply of exportables offered by the "partner" (e_x^* in fig 2.7) and the "rest of the world" (e_x^w in fig. 2.8) at the pre-union terms of trade (tariffs excluded - with tariffs the curves are $\overline{E^t E_x^t}$, $\overline{e^{*t} e_x^{*t}}$ and $\overline{p_1^w e^{wt}}$).

Fig. 2.6.
Trade of the home country

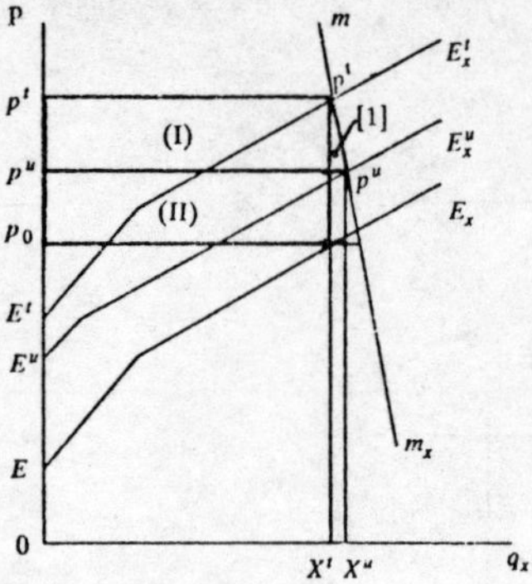

Fig. 2.7
Trade of the partner

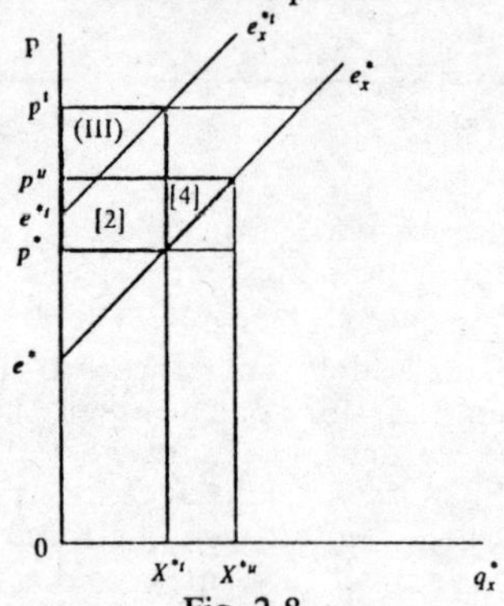

Fig. 2.8
Trade of the rest of the world

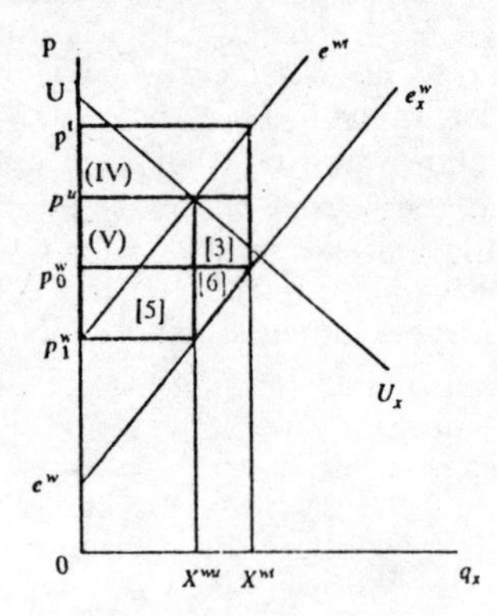

Now, considering that the formation of the customs union brings the terms of trade to p'' (that is, exempting the partner from tariffs and keeping the former[9] tariff as the common customs tariff) we obtain the following results:

(1) Home country gains = [1] + (I)

Production + Consumption effects of trade creation [1]
+
Terms of trade effects (I)

(2) Partner country gains = [2] + [4]

Terms of trade effects on existing exports [2]
+
Trade creation [4]

(3) Rest of the world losses = [5] + [6]

Terms of trade effects for remaining exports [5]
+
Trade diversion [6]

(4) Union net change = [1 + 4 + 5 - 3]

(5) World global welfare effect = [1 + 4 - 3 - 6]

However, the "home country" welfare is also affected by changes in tariff revenue, which in this case are equal to [(I + II) - (V + 5)] = (5 - 2 - 3 - I), so that its net welfare is (1 + 5 - 2 - 3), that is, production plus consumption trade creation effects plus the terms of trade gains relative to the "rest of the world" minus the sum of both trade diversion and the partners terms of trade gain. So, while there is no doubt that the partner would gain from the customs union, one cannot be sure about the home country as this depends on both the terms of trade and trade creation/diversion effects. This clarifies an important point concerning empirical studies of trade creation/diversion effects (even including consumption effects). In other words, the fact that there is net trade creation[10] can only be used to justify the formation of a customs union based on the global gains, but cannot be used to judge whether all individual members benefit from membership.

Now, let us consider the relaxing of the assumption of constant returns to scale. As can be easily understood, one would expect that the member economies could reap economies of scale or location from an enlarged market and stiffer competition. To deal with this question there was an important theoretical contribution by Corden (1972), who built a partial equilibrium model consistent with the simultaneous consideration of trade creation/diversion concepts plus two new concepts measuring economies of scale; these he termed "cost reduction" and "trade suppression". The first occurs when one of the partners captures the whole union market, obtaining a cost reduction effect that is measured by extra profits earned on sales at home. The second effect arises when, due to the existence of the customs union, production is now possible in one of the partners, and replaces cheaper imports from the rest of the world. As with trade diversion, its loss is reflected in the loss of tariff revenue. Nevertheless, as Corden himself pointed out, there is not (and, to our knowledge, this remains so) a satisfactory way of extending these concepts to a general equilibrium model. In the particular context of large unions, which we are interested in, it seems even more difficult to separate out the trade suppression effects from the trade diversion effects.

As we cannot foresee a solution to this problem we shall proceed with an evaluation of the "state of the art" in which we take as unsolved the constant returns to scale assumption.

2.2 - A "requiem" for Customs Union Theory?

In spite of important developments in the theory in the last 35 years (briefly highlighted in the previous section) the impression one gets from the present "state of the art" continues to be very frustrating. The reasons are threefold:

Firstly, in terms of pure positive economics (i.e., regardless of any resemblance to the real world) it is very disappointing that it can not give an answer to the question of the desirability of forming a customs union. Instead, it becomes an empirical matter dependent on the estimation of trade creation and trade diversion effects which are not unequivocally translatable into welfare gains or losses, and it is doubtful if they are measurable at all. Further, according to the argument of Cooper and Massel's (1965) alternative policy (the unilateral non-discriminatory tariff reductions[11]), there is no economic rationale for the formation of a customs union unless one is either prepared to accept the non-economic arguments of the "public goods" type, that there are constraints to the use of first best policies, or to allow for the existence of externalities and market imperfections which extend beyond the boundaries of national states. In either case, we would be outside the general equilibrium "world".

The second major reason for dissatisfaction with the theory stems directly from the fact that general equilibrium economics is criticised for having very little resemblance to the real world economy, and its assumptions, even if taken as heuristic fictions, are considered by some authors to be totally irrelevant. These criticisms form the bulk of the literature, and notable examples can be found in Kaldor (1972,1981). The assumptions were numbered in the previous section and we shall not, here, elaborate upon all the detailed criticisms and the reply from the general equilibrium theorists; we will make only a brief reference to the assumptions we consider to be the most constraining; i.e. the static nature of the theory; the immobility of factors, and the automatic adjustment in the balance of payments.

Starting with the static nature of the theory, this derives mainly from the assumption of given and fixed technology, and not necessarily from the existence of time-lags in the response of economic agents facing tariff changes. This last problem, which is known in the empirical studies as the "once-for-all" assumption about the effects of integration, is further complicated by insufficient consideration of "second round" effects related to substitutability and complementarity of both tradeable and non-tradeable goods and "tertiary effects" via currency movements required by balance of payments imbalance. However, the fundamental issue is that the ignorance of the sectoral and locational investment and reinvestment shifts required by trade reallocation will inevitably change the "given technology", and its direction and characteristics, namely in what concerns both the "embodiment" of technical progress and the Verdoorn relationship[12], which will be the main determinants of the future "revealed comparative advantage" of individual member countries. In this manner, questions like the "infant industry" argument and industrialization strategies can not be accommodated within a simple "public goods" concept, as they are an important determinant of the long run growth performance.

WORLD TRADE MATRIX

	Home country (1)	Partner country (2)	Rest of the world (3)
Home country (1)	-	$x_{12}+c_{12}+d_{32}+r_{12}-s_{12}$	$x_{13}-r_{12}$
Partner country (2)	$x_{21}+c_{21}+d_{31}+r_{21}-s_{21}$	-	$x_{23}-r_{21}$
Rest of the world (3)	$x_{31}-d_{31}+e_{31}-s_{31}$	$x_{32}-d_{32}+e_{32}-s_{32}$	-

with,
c = trade creation between partners
e = external trade creation
d = trade diversion
r = reorientation of exports
s = trade suppression

Also, the assumption of factor immobility at an international level is a severe handicap; this is so primarily because the substitution and complementarity relationships between trade and factor movements are very important and, necessarily, one can always have an "a priori" doubt concerning the consistency of two separate theories - one for customs union and the other for monetary unions (let alone the labour migration theories). Furthermore, direct foreign investment flows are nowadays very large and their location pattern has been significantly affected[13] by commercial policy so that one has also to admit the existence of investment creation and diversion effects, although in their case we cannot be certain whether they are welfare increasing or decreasing.

Finally, the assumption of automatic adjustment in the balance of payments needs careful scrutiny. To start with, it is very unlikely that imbalances arising from tariff changes can be avoided or even taken as negligible in terms of adjustment costs. To illustrate the point we present above a world trade matrix with the trade flows corresponding to an "anti-monde"[14] without integration (x_{ij}) separated from the effects arising from integration.

The balance of trade for each country resultant from the first and (partial[15]) second round effects will be:

$$[(x_{12}-x_{21})+(x_{13}-x_{31})] + [(c_{12}-c_{21})+(d_{32})+(s_{21}+s_{31}-s_{12})-(e_{31})] \qquad 1)$$

$$[(x_{21}-x_{12})+(x_{23}-x_{32})] + [(c_{21}-c_{12})+(d_{31})+(s_{12}+s_{32}-s_{21})-(e_{32})] \qquad 2)$$

$$[(x_{31}-x_{13})+(x_{32}-x_{23})] + [(e_{31}+e_{32})-(d_{31}+d_{32})-(s_{31}+s_{32})] \qquad 3)$$

where the second term in square brackets [] represents the effect induced by integration. It is then easily seen that even if we assume no external trade creation and trade suppression as conventional analytical methods do the union will always face a change in its trade balance as long as there is trade diversion. Furthermore, individual countries as yet unaffected would need to share equally any trade creation gains. This is such a unique combination that it is hard to imagine it occurring other than in the case of identical partner economies. In fact, it would require them to have had a combination of external tariffs and supply curves in the pre-union situation which allowed for the adoption of a common external tariff that did not generate trade diversion, and, in a case where they did not have identical supply and demand schedules, that there should have been a miraculous combination of elasticities to equalise trade creation effects. In short, it is difficult to accept the assumption of no balance of payments effects. Furthermore, in the case of a world made up of large groups such as the EEC, EFTA and USA it is not easy to accept[16] that any changes are so small that they will not induce exchange rate adjustments.

The third reason for dissatisfaction with customs union theory stems from its restriction to trade creation/diversion measures, which require three important qualifications. Firstly, the trade creation/diversion effects do not include the total impact of customs union; secondly, they cannot always be considered as, respectively, welfare increasing or decreasing, and thirdly, it is very unlikely that they can be measured at all.

We will begin by considering Meade's extended analysis of the concepts of trade creation/diversion: this does not include the possibility of trade suppression (Corden 1972) arising from the substitution of cheaper imports from the rest of the world for domestic production. Furthermore, the theory is sensitive to the level of aggregation and if we include a "home sector" we cannot ignore its substitution and complementarity relationship, which can give rise to almost all types of trade flow; i.e., trade creation, trade diversion, trade suppression and external trade creation (trade creation with the rest of the world). Also, as we have shown above, at the union and individual country level the terms of trade effects are important. Finally, there is the possibility of trade reorientation due to a simple "unionist identity" and improved "proximity" due to the free information provided by membership. These problems have long been acknowledged in the literature and some authors have tried to present new taxonomies of effects (eg. Dayal 1977 and Collier 1979).

However, there is no satisfactory way to bypass the Viner - Meade simplifications without having to abandon unequivocal welfare effects. Even within these stricter concepts it is still possible to conceive trade diversion as being welfare increasing instead of decreasing. The situation might arise when diversion of cheaper imports from the rest of the world is partly made by suppressing more expensive, domestically produced substitutes. Several assumptions have been proposed to solve this problem, for instance a suggestion by Johnson (1974), which states that only the diversion of initial trade should be considered and that the new trade is taken as trade creation. Nevertheless, one cannot see a plausible way of separating the two components of a trade flow. Moreover, it may also happen that trade suppression might be welfare increasing. Such a situation (Machlup 1977) arises when as a consequence of the union the partners can set up a production capacity at a higher level of efficiency than the previous "rest of the world" supplier. In particular, the problem with trade suppression effects is that empirically they are hardly (if at all) separable from trade diversion.

Turning now to the last point on the possibility of empirical measurement, we see from the trade matrix above that the net change of individual member trade flows are a combination of different effects. Therefore, as regards imports, the changes in trade for the home country (and also for the partner) are as follows:

imports from the partner = $(c_{21}+d_{31}+r_{21}-s_{21})$

imports from the rest of the world = $(e_{31}-d_{31}-s_{31})$

and regarding exports:

exports to the partner = $(c_{12}+d_{32}+r_{12}-s_{12})$

exports to the rest of the world = $(-r_{12})$

It is now easily seen that using residual methods to estimate changes in trade flows does not allow for the possibility of estimating trade creation (c_{21}) unless one is prepared to accept a world with no external trade creation, with no trade suppression and with only three countries.

The alternative of using the price elasticities of supply and demand, even if we are prepared to forget the old problem of knowing whether price elasticities are different from tariff elasticities (Kreinin 1967) and to accept the "once-for-all" hypothesis, does not provide us with an alternative either. One should be aware of this question and extremely cautious in accepting the various empirical results presented in the vast literature[17] as being true measures of trade creation and diversion effects.

Indeed, until recently the literature has adopted one of the two following approaches: either it takes for granted that whenever the increases in trade flow with the partner are substantially greater than the decrease in trade flow with the third countries then the welfare effects will be positive; or, alternatively, that they are measured by taking half of the volume of increased imports times the difference of the price with and without integration ($\frac{1}{2}(p_t - p_t')[(M_0 - M_0') + (M_P' - M_W)]$ in fig. 2.9) and the volume of imports diverted times the common customs tariff ($[(M_W - M_0)(p_t' - p_w)]$). This is done in accordance with the "small country" assumption and consequent specification of infinite supply curves drawing the third country horizontal line below the partners, which implies (an assumption never clearly stated) that in the pre-union situation the "rest of the world" has the total share of the market. However, this procedure of measurement has been challenged by Tovias (1982) who argues that there is not a one-to-one relationship between the areas corresponding to trade creation and diversion (a + b in fig. 2.9 below) and the estimated change in trade flows (A + B in fig. 2.9). His point is that either we assume that the partner's supply curve is equal to the third country's, in which case there would not be any trade diversion and the welfare effect of trade creation would be measured in the usual way, or that we assume that the partner's supply curve is just inferior by an infinitesimal amount to the home country's and there will be only trade diversion (which in ex-post studies is measured as increased imports from the partner, ie, trade creation). He then argues that estimates should be presented as range values between these two extreme situations.

Nevertheless, although we might agree that the post-integration equilibrium domestic prices (p_t') are rather difficult to obtain this should not be seen as making the correct estimation of trade creation/diversion effects impossible to calculate as long as the assumptions referred to above are acceptable as Tovias' does in his analysis. However, if we drop the assumption that the third country had the total market in the pre-union situation then we have a case where the measurement might not merely be difficult but may even be totally misleading.

Fig. 2.9

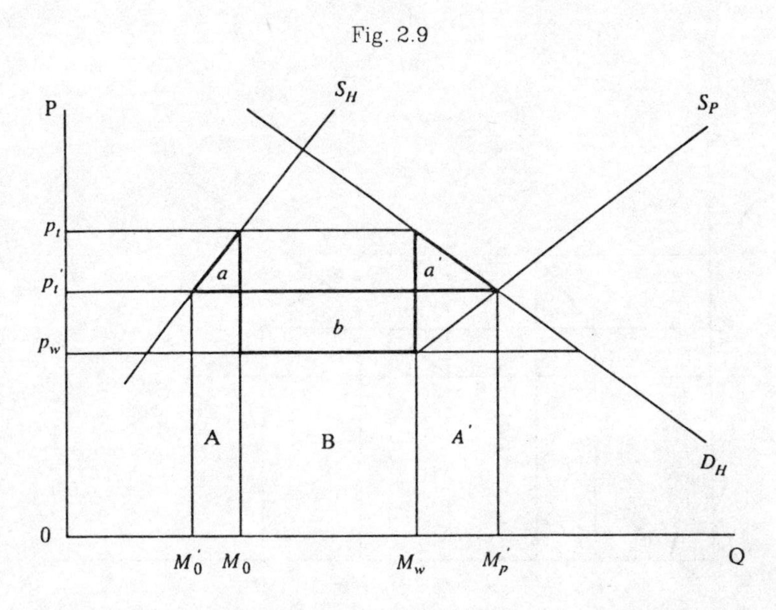

In what might be called the "normal cases", represented in figs. 2.10 and 2.11 below, the partner, in the pre-union situation already has a share (usually the largest) of the market. In fact, empirical studies very seldom are disaggregated at more than one or two digits of the SITC classification and at this level even rather small economies always have trade with both the partner and the third country in the pre - and post - union periods, a fact which rules out the infinite supply elasticity assumption. Furthermore, the partner, which is usually a neighbour, has the advantage over the rest of the world of lower transportation and marketing costs which allows it to make good use of even small markets, although this advantage might be ruled out when larger quantities are involved in which case the rest of the world would step in. Therefore, the partner's supply curve (or its equivalent marginal cost curve) must be seen as increasing after some range of demand, and three situations are possible.

Firstly, the case illustrated by fig. 2.9 above where there is no bilateral trade in the pre-integration period and the post-integration price is such that all external demand can be satisfied by the partner. Here traditional analysis applies except for the share of area a' corresponding to the consumption effect. It is questionable whether this can be labelled trade creation in a Vinerian-Meade sense because an alternative cheaper source of supply is available. This illustrates the point made earlier about customs union formation only being rational as a "second-best" policy as compared with an unilateral non-discriminatory tariff reduction.

Secondly, Fig. 2.10 below illustrates the case where imports from

Fig. 2.10

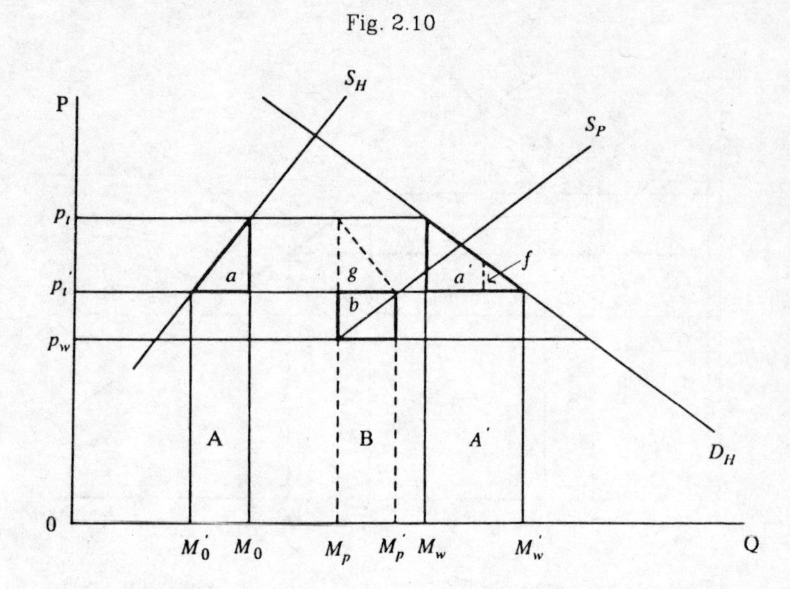

the partner increase ($\overline{M_O'M_O} + \overline{M_PM_P'}$) as well as imports from the rest of the world ($\overline{M_P'M_W'} - \overline{M_PM_W}$), i.e., external trade creation exceeds trade diversion and suppression. However, here the use of the traditional estimation procedure would be misleading, ie, we would get as trade creation ($a+g$), as external trade creation ($a'-g$)[18], and no trade diversion while in fact we should have:

(+) trade creation = a (production effect only)

(-) trade diversion = b

(+) external trade creation = $a' = (M_W' - M_W)(\frac{P_t - P_t'}{2})$

So, although theoretically one can measure total trade creation (i.e. trade creation + external trade creation), provided that we can estimate the change in domestic prices ($p_t - p_t'$), trade diversion (area b) is left out and this effect may be quite substantial. In fact, this effect cannot be measured unless, as in the case of ex-ante studies, we know the exact shape of either the supply curve of the home country or the one of the partner; or, in the case of ex-post studies, we are prepared to accept that there was no change in world prices.

Also, in the case of fig. 2.11 below (that is, in the case where imports from the partner increase but decrease from the rest of the world) the traditional measurement would be totally misleading. It would give as trade creation ($a+g$) and would report[19] as trade diversion the area ($b-d$) while in fact we have:

Fig. 2.11

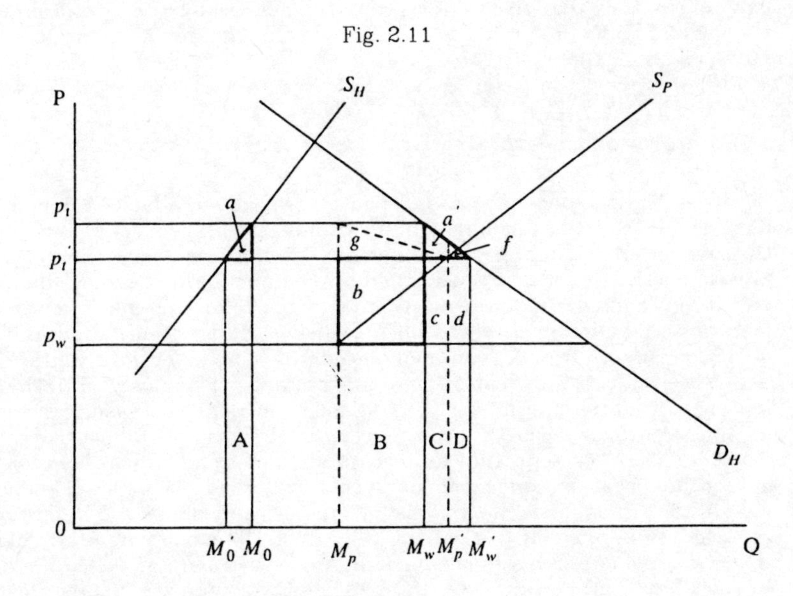

(+) trade creation = $a+(a'-f)$

(-) trade diversion = b

(+) external trade creation = f

Note that in this situation (the more likely if one is to rely on available empirical studies) there could also be a reversing of the conclusions concerning the net effect of integration. In fact, the traditional approach not only underestimates the trade diversion costs but also overestimates the trade creation effects ($g>a'$), as long as there is a relative increase in intra-union trade [i.e., $\frac{(M_P'-M_P)}{(M_W'-M_W)}>1$]. The measurement requirements are the same as in the previous case.

Finally, one can also foresee the case in which there would be no change in total trade with the third country, even though there had been both trade diversion and external trade creation which have not entirely offset each other.

It must therefore be concluded that any net outcome in terms of trade creation/diversion welfare effects is not only subject to bias due to the difficulty of the measurement of price changes, but it is also strongly dependent on the assumed shape of the external supply curves, which raises serious doubts about any conclusions on net welfare effects based on these concepts.

Given these reasons for dissatisfaction with Customs Union Theory, we shall attempt the task of trying to develop an alternative approach to measure integration effects. But, first of all, let us examine other "dynamic" effects not considered by the traditional theory and some of the few attempts to surmount the deficiencies of customs union theory.

2.3 - Integration and growth

One might expect that the ultimate yardstick by which to decide upon the interest of forming/joining an integration process would be the impact on real per capita income of each member. However, almost all the literature[20] has relied on customs union theory which has never considered this question. Indeed, following the Vinerian tradition, it has concentrated on measuring the trade creation and diversion effects which, once calculated, are converted into welfare benefits and costs by estimating the welfare "triangles". But the welfare estimates cannot be seen as an equivalent increase in real income. An example of this can be seen in the case of the estimates made for Spain (Viaene 1982) where, in spite of significant values for net trade creation, once the corresponding trade flows had been linked to a model of the Spanish economy, the simulated results revealed a decrease of 0.3% of the GDP. Also the small magnitude of these welfare estimates [see values in Chapter V , Balassa (1975) and Mayes (1978)] has reinforced the need for protagonists of Customs Unions to rely on expected dynamic effects, which are claimed to be more important than the former. Indeed, such arguments formed the basis of the British decision to join the EEC (see HMSO - 1970) and also that of the Mediterranean countries. It is therefore pertinent to ask what are these "dynamic effects"? Not a single author is exhaustive in this regard, nor do they agree with one another about how these dynamic effects work or about their relative importance.

A complete list of these dynamic effects would include:

(a) The induced investment resulting from the expansion of trade which is measured by assuming a given return to capital and that the share of gross fixed investment in the GNP would have been smaller without integration (Balassa 1975) or by using Denison's method of converting changes in the share of fixed investment in the GDP into increased capital stock which, multiplied by its share of national income, gives the income effect (Krause 1968). Given that the effects of reinvestment induced by income growth due to integration are likely to be rather small, and given reasonable values for the propensity to save and return to capital, the resulting value will obviously be negligible. It is from increased investment and trade flows that the other generally accepted dynamic effects arise.

(b) The efficiency which stems from the expansion of efficient exports and the discouraging of import-competing inefficient activities which are both the result of increased competition and enlarged markets. This efficiency arises from:

economies of scale resulting from the enlarged market which allows the introduction of new efficient activities; longer production runs; concentration by elimination of marginal small-scale activities, and increased specialization.

Although the introduction of new activities is only applicable to small members and the concentration gains are based on the belief that released resources will be used elsewhere, these hypotheses seem consistent even if difficult to quantify.

However, the increased specialisation (probably the most important effect) which results in structural changes in the economy needs some qualifications. Since the nineteen-sixties an increase in intra-industry specialisation[21] has been observed [Balassa (1967,1975); Kreinin (1981)] that has often been ascribed to trade liberalization subsequent to integration (Balassa 1975). However, this simple deduction is rather doubtful given that the same pattern is observed for trade with non-members (Kreinin 1981), and it seems to exist everywhere except in the very low-wage countries (Glejser-Goossens-Vanden Eede 1982). Furthermore, in an attempt to explain this pattern it was found (Caves 1981) that while scale economies had a significant effect, only weak statistical support was found for the variance of countries' tariff rates and none for their level.

Alternatively, if we accept that integration has an important role in this process then the possibility that resources move from import-competing to export industries is reduced. This implies that member countries will become more similar and that the cost of any adjustment and changes in income distribution that might result from integration is substantially reduced.

Intra-industry specialization also enhances the likely importance of increased competition so that we might expect that X-efficiency gains are to increase as a consequence of integration. This concept, developed by Leibenstein (1966), relies on the idea that entrepreneurs face an indifference curve between profits and leisure (or lethargy). So, the removal of profits will produce both income effects, which induce greater managerial effort, and substitution effects inducing increased leisure; the global outcome being dependent upon the relative importance of the two effects.

Some authors, [Williamson (1971); Balassa (1975)], support the idea that these gains are likely to be significant because the "representative firms" produce both exports and import-substitutes so that customs union entry would result in a steeper transformation curve intersecting the old one at the previous equilibrium point. Williamson even suggests that the superior managerial effort gains could be greater than the conventional gains from trade on the assumption that nominal X-efficiency gains (say 1%) could be obtained over a larger proportion of GNP (say 40%, which would result in a 0.4% of GNP gain). However, the fact that these gains cannot be entirely described as welfare benefits due to disruption of leisure or mere redistribution effects, and the fact that there are no reliable measurements[22], has prevented most authors from considering them.

The learning-by-doing gains are also imputed to arise as a consequence of integration through the effect of larger production runs and "cultural+information" flows. However, there are no quantitative estimates and all the projections are of a hypothetical nature, such as in the case of: transport cost gains which are expected to result from improved common infrastructures; common policies, and an expansion of trade between contiguous regions [Nugent (1974) and Erlenkotter (1972)].

We therefore conclude that the sources of efficiency gains are both difficult to isolate and to quantify. The few aggregate measurements suggested are the ones by Balassa (1975) and Krause (1968). The first uses an estimate of returns to scale of 1.3 indicated by Walters (1963) (which is above the normal values found in other production function studies) so that the estimated new trade of around 2.3% of GNP implies a gain in GNP of slightly over 0.5%. The second uses the increment in ratios of imports to GDP which, when multiplied by the corresponding import tariffs, gives the cost savings from greater efficiency (the result is annualised for comparison). These are rather crude estimates which fail to distinguish between the sources of increased efficiency strictly dependent on the trade effects of the customs union, and which rely on gross estimates[23].

(c) The terms of trade effects have also been considered by some studies [Williamson (1971) ; Miller-Spencer (1976); Viaene (1982); Miller (1971), and Petith (1977)] as being a significant outcome of integration. Indeed, it must be remembered from the previous section that the terms of trade can be dealt with within the Customs Union Theory and should not be strictly defined as "dynamic effects". Nevertheless, the problem is that it cannot be established "a priori" what sign the effect will have, as this depends on the proposed tariff changes and their level; on whether we have complementary or substitute goods; on the size of the member countries, and the possibility of retaliation by non-member countries.

However, the few studies that present direct estimates (based on general equilibrium models) rely on oversimplistic assumptions regarding the number of countries, constant returns to scale and elasticities. Not surprisingly, the results vary from very low values for Spain (+) (Viaene 1982) and the UK(-) (Miller-Spencer 1976) to quite substantial values (+) for Germany and France (Petith 1977). Petith claims that the differences are due to the value we assume for the ratio of the elasticity of substitution of imports themselves to the elasticity of substitution of domestic production and imports. He also is the only one to convert them in terms of income values achieving average results between 0.34% and 0.93% of GNP (depending on the value of the ratio between elasticities of substitution with values ranging from 2.4 to 8.8).

If these values were to be relied upon then the terms of trade effects would be shown to be the single most important factor of economic integration. Therefore, both their neglect or the simple estimation of the terms of trade change required for restoration of the balance of payments using the Marshall-Lerner condition are highly questionable procedures.

(d) The balance of payments impact of integration has in itself been regarded as another dynamic benefit/cost of integration. But, here also, there is no clear explanation of the mechanisms by which it influences income growth. The importance of these effects has mostly been seen in the context of introducing the implications of the agricultural policy and budget contributions [see Kaldor (1971); Miller (1971) and Penketh (1982)].

(e) Finally there is the question of whether or not the wage-price effects should be taken account of and, if so, whether on the credit or debit side. Although most studies have ignored them, the Kaldor and Miller articles referred to above have dealt with them as part of the so-called resource cost of membership.

It is not difficult to see that all the attempts to measure the growth effects of integration referred to above suffer from severe limitations. This is so because of the rather crude nature of the calculations and assumptions and because of the partial nature of the studies. In fact, they do not come to terms, in any measurable way, with the inadequacy of estimates of trade creation and diversion based on customs union theory.

However, although the bulk of the literature is based on customs union theory, two other alternatives have been proposed. The first, suggested by Kaldor (1971), involves complementing the estimated trade creation/diversion effects with the so-called resource costs/benefits of membership made up of net contributions to the budget; the increased costs[24], and net capital flows and costs of adjusting the balance of payments, against which one should analyse the conjectured expected dynamic benefits/costs [see Kaldor (1971), Miller (1971) and Penketh (1982) for such estimates for the U.K.]. The second approach [Nugent (1974)], used a Cobb-Douglas production function with two dummy variables to pick up "once-for-all" and gradual effects of integration on the per-capita income. This, as the author acknowledges, is a very crude approach which is extremely sensitive to the various specifications used. The resource cost/benefit approach is also very crude in its nature and does not provide any way of converting the measured effects into income growth.

We conclude that up until now there has been no satisfactory method developed for measuring the integration effects on economic growth and income. We can summarise as follows:

Firstly - customs union theory (even when taken as a strict contribution to positive economics) is in a deadlock.

Secondly - the reliance of customs union theory on the trade creation and trade diversion effects introduces the virtually insurmountable problem of the empirical measurement of these effects.

Thirdly - customs union theory excludes the consideration of the so-called "dynamic effects" for which no satisfactory method of measurement exists.

Fourthly - in the EEC case it fails to consider one of the common market's basic features: the free mobility of factors (capital and labour).

Fifthly - it neglects the institutional and policy set up (eg. common budget, agricultural, industrial and regional policies, EMS, etc.) which is a serious deficiency.

Finally - the suggested alternative methods are too aggregative to be considered as useful alternatives.

We will attempt here to provide a more fruitful methodology for measuring integration effects on growth.

1. From now on we will number in square brackets [] all the assumptions being used in this section.

2. According to this hypothesis the dismantling of tariffs between members may cause a reallocation of direct foreign investment of the members in favour of non-members, as well as an increased flow of investment from the "rest of the world" into the union.

3. Theoretically, at this stage, there is no reason why we could not also consider the cases of increasing or decreasing returns to scale, but by redrawing figure 2.3 accordingly one can immediately see that there would not exist very much scope for a customs union.

4. In fact changes in the terms of trade (p) will result in movements along the contract curve, therefore affecting relative factor incomes, but conventionally trade and tariff theory abstract from distribution considerations.

5. Model builders in the general equilibrium tradition have surmounted this problem by representing any imbalances by a terms of trade change between the tradeable and the non-tradeable sectors. However, it is not clear that we can use the same device in this context.

6. In this regard there is also an important separation concerning the case of less developed economies where questions of industrialization and other issues are paramount. See Robson (1983) on the treatment of this special case.

7. It should be remembered that the free trade area EEC-EFTA represents the largest market in the world with more than 310 million inhabitants.

8. This has been traditionally used by considering that the home country is so "small" that the customs union does not affect the terms of trade in either of the two economies.

9. This is the home country optimal unilateral tariff in the case where discriminatory tariffs can not be used.

10. Even admitting that actual empirical studies can separate trade creation effects, which we challenge in the next section.

11. To a level identical to the one resulting from membership but without any tariff revenue loss.

12. It should be remembered that most studies of production functions ("Nelson type") have shown that the rate of growth attributed to "embodied technical progress", is very large, and McCombie-Ridder (1984) have shown the existence of substantial inter-industry differences for the Verdoorn relationship between productivity growth and output growth.

13. Later in chapter V we shall present evidence in favour of the so-called tariff discrimination hypothesis.

14. An "anti-monde" is defined as the situation that a given economy would (or will) have if everything else had evolved without being affected by the change in consideration (integration).

15. A complete account is not possible even in general equilibrium models unless the home market (tradeable and non-tradeable) is also included and home income effects are taken into account.

16. Results presented in chapter V confirm this point.

17. A summary of these results is presented in chapter IV.

18. Note, that this is so because empirical studies usually only measure net changes in imports from the "rest of the world" rather than just external trade creation (i.e., external trade creation minus trade diversion and suppression), which in Fig. 2.10 corresponds to the base of triangle f, which is multiplied by half of the change in domestic prices ($p_t - p_t'$) instead of its height so that one ends up with a total welfare measure of the consumption effects equal to a'. Indeed, if we were using the area of triangle f as the true measurement of external trade creation then a' could be greater, equal or smaller than ($g + f$), depending on the relative increases of imports from the "partner" and the "rest of the world" [ie, $a' \gtreqless (g+f)$ iff: $\frac{(M_P' - M_P)}{(M_W' - M_W)} \lesseqgtr 1$].

19. See previous footnote.

20. Two review papers suggested are the ones by Verdoorn-Bochove (1972) and Mayes (1978).

21. The most used measure is the ratio $\frac{1}{m}\sum_{i=1}^{m}\frac{|X_i - M_i|}{X_i + M_i}$ which takes a value closer to zero the greater the intra-industry specialization and where X_i, M_i are respectively the exports/imports of commodity i and m is the number of commodity categories considered.

22. See Nugent (1974) for a suggestion based on the identification of commodities which would be produced domestically.

23. In the balance of payments constrained growth approach developed in Chapter III no direct measure was attempted as they are included in some of the variables used, such as the terms of trade.

24. The item usually considered in the EEC case is the increased price of food due to the operation of the Common Agricultural Policy.

Chapter Three

AN ALTERNATIVE APPROACH TO CUSTOMS UNION THEORY

3.1 - Introduction

Customs union theory is a well-established field in economics and its assumptions and predictions have been extensively discussed. Good surveys have been done by Lipsey (1960), Johnson (1965), Krauss (1972) and Jones and El-Agraa (1981). Previously (section 2.3) we added our own views on its main assumptions. We can summarise here the main drawbacks. These derive from its general equilibrium core assumptions, namely that: there are no factor movements; there is an automatic adjustment of the balance of payments, and technology is fixed and constant returns to scale are assumed. Customs union theory then relies on the Vinerian-Meade concepts of trade creation and trade diversion which, although theoretically important, do not exhaust all integration effects. Further attempts to introduce terms of trade changes and economies of scale are mostly of a partial and very restrictive nature.

Apart from contradictions in these concepts [Johnson (1974)] when related to welfare effects, and the difficulties of empirical measurement [Tovias (1982)], our point of departure here is that the trade creation and trade diversion concepts cannot appropriately be used to measure growth effects. Not only are static concepts ill-suited to dynamic measurement, but (i) the balance of payments does not automatically adjust; (ii) there are a few cases when trade creation and trade diversion cannot be empirically measured, and (iii) in the EEC case they would be incomplete due to the existence of Common Market policies.

Here we take the view that total trade effects must be considered in evaluating integration, regardless of whether these come about through trade creation/diversion, external trade creation, trade reorientation or trade suppression. If it is accepted that integration effects are mainly derived from trade then the most fruitful procedure for measurement is to use the foreign trade multiplier, with export growth as the major component of autonomous demand.

The advantage of using a multiplier is self-evident if one considers that the traditional customs union effects are of a static nature while most of the literature [Williamson (1971), Balassa (1975), Mayes (1978;1983) and many others] claims that the dynamic effects are likely to be at least of the same size or even greater. However, apart from the earlier work of Brown (1961), who first exemplified the effects of integration by means of the investment multiplier, no other work has been done using this approach. Only recently, El-Agraa (1979) presented a three-country version of Brown's model but still using the investment multiplier while (Begg et al. 1981) used the multiplier applied to the energy sector.

Here we shall use and modify the framework outlined by Thirlwall in a series of recent papers (1979;1982): a framework that can be traced back to Harrod (1933).

3.2 - The Balance of Payments Constrained Growth Model Framework

Thirlwall's main contention is that the balance of payments position sets the limit to the growth of demand to which supply can adapt, and that therefore the long-run growth rate can be approximated by the so-called dynamic version of the foreign trade multiplier[1]:

$$y_B = \frac{x}{\pi}$$

where y_B is growth rate of output; x is growth rate of export volume and π is income elasticity of demand for imports. It has also been shown (McCombie 1985) that this "Thirlwall's law " is identical to the Hicks super-multiplier if the growth of exports and other autonomous expenditures are the same. Thirlwall (1979, 1982) has also provided statistical evidence in favour of his theory for both developed and developing economies and has examined the conditions of divergence from his law due to sustained capital outflows or inflows and relative price movements in international trade. A relation between the growth rate allowed by the trade multiplier, the growth rate with a disequilibrium balance of payments and the actual growth rate can then be derived.

We start from the basic national accounts ex-post identity:

$$P_d Y = P_d A + P_d X - P_f EM \qquad 3.1)$$

where P_d is index of domestic prices; Y is national income; A is domestic absorption; P_f is index of foreign prices; E is the exchange rate measured as the domestic price of foreign currencies; M is the volume of imports, and X is the volume of exports.

In real terms:

$$Y = A + X - \frac{P_f E}{P_d} M \tag{3.2}$$

Thus any excess of real expenditure over real income will have to be met by real capital inflows (C); that is:

$$X - \frac{P_f E}{P_d} M + C = 0 \tag{3.3}$$

Alternatively, we can write[2]:

$$P_d X + P_d C = P_f EM \tag{3.4}$$

and by taking rates of change we obtain,

$$\frac{X}{X+C} x + \frac{C}{X+C} c + p_d = p_f + m + e \tag{3.5}$$

where the lower case letters represent the growth rates.

Now specify the import and export functions as

$$M = \left(\frac{P_f E}{P_d}\right)^{\psi} Y^{\pi} \tag{3.6}$$

and

$$X = \left(\frac{P_d}{P_f E}\right)^{\eta} Z^{\epsilon} \tag{3.7}$$

where ψ is price elasticity of demand for imports; π is income elasticity of demand for imports; η is price elasticity of demand for exports; Z is the world income, and ϵ is income elasticity of demand for exports. Then taking rates of change, gives:

$$m = \psi(p_f + e - p_d) + \pi(y) \tag{3.8}$$

and

$$x = \eta(p_d - p_f - e) + \epsilon(z) \tag{3.9}$$

Substituting (3.8) and (3.9) in expression (3.5) above gives the growth rate consistent with initial disequilibrium in the balance of payments, that is,

$$y = \frac{\left(\frac{X}{X + C}\eta + \psi\right)(p_d - p_f - e) + (p_d - p_f - e) + \frac{X}{X+C}\epsilon(z) + \frac{C}{X+C}c}{\pi} \tag{3.10}$$

In the numerator we have now distinguished the volume effect of relative price changes (first term), the pure terms of trade effect $(p_d - p_f - e)$, the exogenously determined growth of exports and the effect of capital movements.

If we now assume that the terms of trade remain unchanged, that is,

$p_d = e + p_f$, the identity reduces to a more manageable form:

$$y_B^* = \frac{\frac{X}{X+C}x + \frac{C}{X+C}c}{\pi} \qquad 3.11)$$

However, this expression is not exempt from terms of trade and volume price effects and no matter how close the trade multiplier growth rate approaches the actual growth rate one cannot claim the validity of "Thirlwall's law" based on this relationship[3].

Alternatively (Mendes 1985), we can rewrite 3.11) above, as:

$$y_B^* = y\left(1 + \frac{g}{m}\right) \qquad 3.12)$$

where g is the change in the terms of trade [$-(p_d - p_f - e)$] and m is the growth rate of imports. This specification subtracts both export and import price volume effects plus an interdependence term between imports and the terms of trade. This allows an easier decomposition of the actual growth rate into the various components of growth: the trade multiplier, capital flows, terms of trade and price volume effects.

Further, it can be shown that the closeness of "Thirlwall's law" to the actual growth rate is dependent on the growth changes induced by capital flows and terms of trade changes offsetting each other and on aggregate import price elasticities being generally low. As empirical evidence[4] supports this last hypothesis one would expect very good predictions of the actual growth rate from this accounting framework.

In fact, we have found (Mendes 1985) evidence for 12 industrialized economies which only presented an average absolute difference of .31 % points and showed that in the period 1968-81 the trade multiplier could "explain", on average, 80% of the growth experienced by western developed economies. In table 3.1 above we present further evidence covering the 1960-72 and 1973-81 periods. In spite of invisibles and trade services not having been taken into account, the closeness between estimated and actual growth rates is very high (with a mean absolute difference of 1.16 and .38 % points for the first and second periods respectively).

During both periods the trade multiplier growth rate was very close (around 80%) to the actual value although the contributions of capital flows and movements to the terms of trade were reversed. In the 1960's all countries experienced trade surpluses (capital outflows) and improved their terms of trade, while during the 1970's all (except the U.K.) experienced a worsening of the terms of trade. It can also be verified that in both periods the price volume effects were small.

We can now extend and modify this balance of payments framework to analyse integration effects.

Table 3.1 - Sources of Growth in European Economies

Countries	Actual Growth Rate [1 = 2+3-4-5] (1)	Sources of Growth			
		y_B Trade Multiplier (2)	$(y_B^* - y_B)$ Capital Flows (3)	$\frac{g}{\pi}$ Terms of Trade (4)	$\frac{g}{\pi}\psi(-\frac{g}{m})$ Price Volume Effects (5)
Period 1960-72					
Germany	4.40	7.86	-4.89	-2.64	1.21
France	5.40	5.42	-1.12	-1.29	0.19
Italy	5.10	5.31	-1.75	-1.37	-0.17
Netherlands	5.10	5.47	-2.99	-2.80	0.19
Belgium-Lux.	4.60	5.22	-1.59	-1.03	0.06
Austria	4.70	4.98	-0.98	-0.75	0.05
Finland	4.40	4.60	-0.65	-0.45	0.00
Norway	4.10	4.62	-1.68	-1.49	0.29
Portugal	6.30	6.78	-1.06	-0.57	0.04
Sweden	4.00	4.49	-1.47	-1.06	0.03
Switzerland	4.20	3.39	-0.37	-0.99	-0.18
U.K.	2.80	3.00	-0.88	-0.64	0.00
Denmark	4.20	5.83	-3.42	-2.80	0.98
Period 1973-81					
Germany	2.40	2.17	0.67	0.50	-0.06
France	2.70	3.26	-0.18	0.39	-0.02
Italy	2.60	3.46	-0.55	0.30	0.01
Netherlands	2.10	1.91	0.27	0.06	0.01
Belgium-Lux.	2.10	2.17	0.56	0.62	0.01
U.Kingdom	1.10	1.56	-2.33	-1.31	-0.56
Ireland	3.80	3.45	1.58	0.86	0.37
Denmark	1.80	3.01	0.45	0.85	0.81
Austria	2.80	3.62	-0.87	0.00	0.00
Iceland	3.50	5.77	-2.26	-0.02	-0.01
Finland	2.60	3.92	2.09	1.49	1.90
Norway	4.40	3.01	2.67	0.27	1.04
Portugal	3.50	3.22	4.70	2.21	2.24
Sweden	1.40	1.37	0.60	0.46	0.11
Switzerland	4.90	2.01	1.84	-0.52	-0.56

Note: a negative sign in the terms of trade means an improvement.

3.3 - The Measurement of Integration Effects within the Balance of Payments Framework

We start with the overall balance of payments identity in equation (3.3):

$$X-\left(\frac{P_f E}{P_d}\right)M+C=0 \qquad 3.12)$$

Now let $M=KY$ where K is the average propensity to import so that:

$$Y=\left(\frac{X+C}{K}\right)\left(\frac{P_d}{P_f E}\right) \qquad 3.13)$$

Taking rates of growth gives:

$$y=(X+C)-k+(p_d-p_f-e) \qquad 3.14)$$

That is, the growth rate equals the rate of growth of exports and capital flows $(X+C)$ minus the growth of the propensity to import (k) plus the growth of the real terms of trade (p_d-p_f-e). The change in the growth rate induced by integration will be equal to the induced changes in these components, that is,

$$\Delta y=\Delta(X+C)-\Delta k+\Delta(p_d-p_f-e) \qquad 3.15)$$

Now rewriting 3.12) as,

$$X+C=GM \qquad 3.16)$$

where $G=\frac{P_f E}{P_d}$ so that $g=(p_f+e-p_d)=-(p_d-p_f-e)$

we have,

$$(X+C)=\frac{1}{GM}\frac{d(GM)}{dt} \qquad 3.17)$$

or

$$(X+C)=\frac{MdG}{GMdt}+\frac{GdM}{GMdt} \qquad 3.18)$$

and

$$(X+C)=g+m \qquad 3.19)$$

so that 3.14) and 3.15) above can be rewritten as

$$y=m-k \qquad 3.20)$$

and

$$\Delta y=\Delta m-\Delta k \qquad 3.21)$$

That is, the growth induced by integration will equal the change in the growth rate of imports minus the change in the growth of the propensity to import.

Given the estimates (through the share technique) of the annual volume of imports induced by integration, the change in import growth can be easily estimated.

Since

$$M = M' + I \qquad 3.22)$$

where M' is the import volume without EEC and I is import volume generated by the EEC[5], the change in the growth of imports can be represented as follows:

given that,

$$m' = m\frac{M}{M-I} - i\frac{I}{M-I} \qquad 3.23)$$

where i is the growth rate of I and since $\Delta m = m - m'$ we have:

$$\Delta m = m(1 - \frac{M}{M-I}) + i\frac{I}{M-I} \qquad 3.24)$$

or

$$\Delta m = \frac{I}{M-I}(i-m) \qquad 3.25)$$

Finally, the change in growth of the propensity to import (Δk) can be estimated as follows:

$$K = \frac{M}{Y} \qquad 3.26)$$

so that

$$k = \frac{1}{M}\frac{dM}{dt} - \frac{1}{Y}\frac{dY}{dt} \qquad 3.27)$$

and

$$k\frac{Mdt}{dM} = 1 - \frac{dY}{dM}\frac{M}{Y} \qquad 3.28)$$

or

$$k = (1 - \frac{1}{\pi})m \qquad 3.29)$$

Therefore,

$$\Delta k = k - k' = m - \frac{m}{\pi} - m' + \frac{m'}{\pi'} \qquad 3.30)$$

and

$$\Delta k = \Delta m + \frac{m\Delta\pi - \pi\Delta m}{\pi(\pi - \Delta\pi)} \qquad 3.31)$$

To estimate Δk we need to measure the impact of the EEC upon the income elasticity of demand for imports of its members ($\Delta\pi$). This will come about through changes in trade specialization and there are no clear "a priori" expectations regarding its magnitude and sign[6]. Therefore, its evaluation is an empirical matter.

Given reliable estimates of the EEC effect on the growth rate of imports and in the income elasticity of demand for imports, we can have an estimate of the overall effect upon the growth rate, as:

$$\Delta y = \frac{\pi \Delta m - m \Delta \pi}{\pi(\pi - \Delta \pi)} \qquad 3.32)$$

We can see, then, that the sign and magnitude of the integration effects will be dependent on the integration induced imports (I); on the difference between i and m rates of import growth ($i-m$), and on the induced change in the income elasticity of demand for imports ($\Delta\pi$).

It is important to remember from the previous chapter that the total effect of integration is equal to the difference between trade creation and reorientation minus trade suppression, i.e.,

$$I = [(c_{21} + e_{31}) + r_{21} - (s_{21} + s_{31})] \qquad 3.33)$$

Even if there is net trade creation[7] ($I>0$), this condition is not sufficient to have a positive impact on growth as conventional analysis assumes. Furthermore, if we were also to adopt the traditional hypothesis of "once-for-all" effects (ie, i = 0), then a positive contribution to growth would require a reduction of the income elasticity of demand for imports [$(\Delta\pi < -\pi(\frac{I}{M-I})$].

It should also be noted that our model shows that the occurrence of net trade creation for individual members is not a necessary condition for a positive contribution to growth. In fact, there are several possibilities of obtaining a positive contribution depending on the values of i and $\Delta\pi$.

However, and perhaps more important, one of the possibilities of our model is that the effect on the growth rate can be split up into its component parts in order to assess the different transmission mechanisms and check the corresponding results against the theoretical and empirical evidence available. This will be done in the following way:

First, the effect on the terms of trade $[\Delta(p_d - p_f - e)]$ can be derived from the balance of payments growth rate with a constant terms of trade [y_B^* - see previous section].

Rewriting 3.12) above as,

$$g = \frac{m}{y}(y_B^* - y) \qquad 3.34)$$

and without integration,

$$g' = \frac{m'}{y'}(y_B^{*\prime} - y') \qquad 3.35)$$

we have,

$$\Delta g = g - g' = \frac{m}{y}(y_B^* - y) - \frac{m'}{y'}(y_B^{*\prime} - y') \qquad 3.36)$$

And, as

$$m' = m - \Delta m$$

$$y_B^{*\prime} = y_B^* - \Delta y_B^*$$

$$y' = y - \Delta y$$

we can rewrite 3.36) above, as

$$\Delta g = \frac{m}{y}y_B^* - \frac{m}{y}y - \left(\frac{m - \Delta m}{y - \Delta y}\right)(y_B^* - \Delta y_B^* - y + \Delta y) \qquad 3.37)$$

which, after transformation, becomes:

$$\Delta g = \frac{y\,\Delta m\,(y_B^* - \Delta y_B^*) + y\,\Delta m\,(\Delta y - y) + m\,(y\,\Delta y_B^* - y_B^*\Delta y)}{y(y - \Delta y)} \qquad 3.38)$$

or

$$\Delta g = \frac{y\,\Delta m\,(y_B^* - \Delta y_B^* + \Delta y - y) + m\,(y\,\Delta y_B^* - y_B^*\Delta y)}{y(y - \Delta y)} \qquad 3.39)$$

This will require an estimation of the change induced in the balance of payments growth rate (Δy_B^*) which can be estimated (under reasonable assumptions) once we know the integration effect on trade (exports) and on the income elasticity ($\Delta\pi$).

From 3.11) in the previous section we have,

$$y_B^* = \frac{\frac{X}{X+C}x + \frac{C}{X+C}c}{\pi} \qquad 3.40)$$

so that

$$y_B^{*\prime} = \frac{\frac{(Xx + Cc) - (Ee + Ff)}{(X + C) - (E + F)}}{\pi - \Delta\pi} \qquad 3.41)$$

where E is the export volume induced by the EEC; e is the export growth induced by the EEC; F are the financial flows induced by EEC, and f is the growth of financial flows induced by EEC.

The export volume effect (E), is estimated through the share technique and the capital flows effect (F), is estimated from the equilibrium condition of expression 3.16) above.

Note, however, that

$$F = I(G - \Delta G) + M\,\Delta G - E \qquad 3.42)$$

i.e., an interdependence problem arises from the inclusion of the change in the terms of trade in the estimation of the capital flows (F).

An alternative method to estimate the changes in the terms of trade is the use of expression 3.19) above, as

$$\Delta g = \Delta(X+C) - \Delta m \qquad 3.43)$$

in which case the capital flows would be split up into its components, that is: where F_1 is change in current balance due to trade flows and unaccounted effects;

F_2 is labour remittances;

F_3 is direct foreign investment, and

F_4 is net budget payments.

Given that F_2, F_3 and F_4 can be independently estimated and assumed not to be influenced by changes in the terms of trade, we can estimate the change in the current balance position as,

$$F_1 = I(G-\Delta G) + M\Delta G - E - F_2 - F_3 - F_4 \qquad 3.44)$$

but, we still are left with the interdependence problem.

Therefore it follows that the estimation[8] of the change in the terms of trade growth rate will only imply the assumption that $y_B^{*\prime}$ will not be substantially affected by terms of trade effects on the induced growth of exports and capital flows.

Or alternatively one can take only the trade balance, say

$$F_1 = I - E \qquad 3.45)$$

so that we are left with only an interdependence factor, which is:

$$\Delta G(I-M) = I(G-1) - (F_2+F_3+F_4) \qquad 3.46)$$

Now, given the estimation of the pure terms of trade effect of integration, we can estimate the change due to exports and capital flows $[\Delta(X+C)]$, and split it up.

From 3.15) we have,

$$\Delta(X+C) = \Delta y - \Delta(p_d - p_f - e) + \Delta k \qquad 3.47)$$

which can be rewritten as,

$$\Delta(X+C) = (X+C) - (X+C)' \qquad 3.48)$$

and

$$\Delta(X+C) = \frac{1}{X+C}(Xx+Cc) - \frac{1}{X'+C'}(X'x'+C'c') \qquad 3.49)$$

Therefore, with $X' = X - E$ and $C' = C - F_1 - F_2 - \cdots - F_n$ where E is exports induced by integration and F is capital flows induced by integration (budget, labour remittances, private investment, etc.), we can transform expression 3.49) by substitution of the growth rates x' and c' so that,

$$\Delta(X+C)=\frac{(Xx+Cc)}{X+C}-\frac{(Xx-Ee+Cc-F_1f_1-\cdots-F_nf_n)}{(X+C)-(E+F_1+\cdots+F_n)} \qquad 3.50)$$

and

$$\Delta(X+C)=\frac{(Ee+F_1f_1+\cdots+F_nf_n)-(Xx+Cc)\left(\frac{E+F_1+\cdots+F_n}{X+C}\right)}{(X+C)-(E+F_1+\cdots+F_n)} \qquad 3.51)$$

Using 3.51) we can now estimate each contribution to $\Delta(X+C)$ such as obtained from 3.47), by successive adding up of estimated capital flows[9] attributable to integration.

We can now provide country (or global) estimates of the growth effects of integration. Furthermore, these can be split up into components that have been widely recognized in the literature as being the most important, that is, trade effects; terms of trade, and allocation of production factors and transfers.

Therefore, in the next chapters empirical evidence disaggregated by transmission mechanism will be given for the effects of the EEC upon individual members, including:

(e) - growth of export volume
[estimated from expression 3.51]

(f_1) - change in the trade balance position
[estimated from expression 3.51]

(Δk) - change in the propensity to import
$[\ \Delta k=\Delta m+\frac{m\,\Delta\pi-\pi\Delta m}{\pi(\pi-\Delta\pi)}\ ]$

(Δg) - terms of trade changes
$[\ \Delta g=\frac{y\,\Delta m\,(y_B^*-\Delta y_B^*+\Delta y-y)+m\,(y\,\Delta y_B^*-y_B^*\Delta y)}{y(y-\Delta y)}\ ]$

(f_2) - labour remittances
[estimated from expression 3.51]

(f_3) - foreign investment
[estimated from expression 3.51]

(f_4) - net budget payments
[estimated from expression 3.51]

We can summarise the main features of this new balance of payments framework as follows:

Firstly - since no country can grow at a faster rate than that rate consistent with balance of payments equilibrium on current account, unless it can finance deficits, and as every country will have a growth rate consistent with overall balance, the new framework relies on using the foreign trade multiplier in its dynamic version which is better-suited than the static approach used by customs union theory.

Secondly - the new framework, instead of the trade creation/diversion and welfare measures, uses total trade effects to estimate changes in output which are both simpler and more correct.

Thirdly - the new framework also takes into account the import side through changes in the income elasticity of demand for imports. This is particularly important because we cannot expect an automatic adjustment of the balance of payments.

Fourthly - it also shows that the occurrence of net positive trade creation effects is neither a necessary nor a sufficient condition to obtain an increase in output.

Finally - the new framework accounts in an integrated way for the most important effects of integration, namely: trade effects, terms of trade changes and factor mobility.

We proceed with developments of the weighted share technique used to estimate the trade effects of integration needed for this new framework.

1. Harrod's foreign trade multiplier is $\frac{\Delta Y}{\Delta X} = \frac{1}{n}$ where n is the marginal propensity to import (Harrod 1933).

2. Thirlwall expresses it as $P_d X + C = P_f M$, where C represents capital flows in nominal terms, which leads to a more complicated expression for the disequilibrium balance of payments growth rate $[y_B^* = \frac{\frac{P_d X}{P_D X + C} + \frac{C}{P_D X + C}(c - p_d)}{\pi}]$. This can be seen as collapsing to our derivation by simply substitution of C by $P_d C$ and some algebraic manipulation.

3. x will include the effect of price changes if they have occurred.

4. See Stern et al (1976); Thursby (1984) and Mendes (1985a) for empirical evidence on price elasticities.

5. Note that from here on $'$ will refer to a without integration situation.

6. Balassa's (1967) analysis in terms of changes in the income elasticities would probably tend to suggest an increase given the high level of intra-EEC trade.

7. Here we use the change in total imports (I) due to integration as the closest measure of the true trade creation effects, because it also includes external trade creation and only requires the assumption that trade reorientation and trade suppression effects are either nil or that they cancel out. The alternative of using changes in imports from the partner countries followed by most empirical studies requires a more strict version of this same assumption while disregarding external trade creation. Furthermore, this alternative still needs to presume that trade diversion effect can be measured empirically or that they are negligible, which is very doubtful.

8. Two alternative ways of estimating the change in the terms of trade would be: 1 - by assuming that the relation between the multiplier growth rate (y_B) and the growth rate allowed by the balance of payments position (y_B^*) would be the same with and without integration (that is, $y_B^* = \alpha y_B$ and $y_B^{*\prime} = \alpha y_B{}'$). 2 - reestimation of the export function for the post-integration period using the residual technique. Both these methods can be used to check the reliability of the estimates.

9. Similarly to what was said above about expression 3.32, here the existence of creation effects, say investment creation, is not a sufficient or necessary condition to obtain a positive effect on the growth rate. Indeed, we cannot even foresee the corresponding sign of a particular flow because this depends also on the relative magnitude of the numerator and denominator, that is, of the value of all the other effects.

Chapter Four

ESTIMATING INTEGRATION EFFECTS ON TRADE

4.1 - A brief survey

"On ne se baigne deux fois dans le meme fleuve"
Heraclitus 500 b.c.

It will become evident in this chapter how relevant is the quotation above and how it sums up all attempts to measure the effects of European integration. In fact, the construction of an "anti-monde", that is, a model based upon what would have occurred in the absence of integration, on which all the trade effects are based, is an impossible task. All measurement therefore requires a certain amount of "faith" and the most careful evaluation of every figure.

There are at the moment more than twenty-five estimates regarding the global effect of the EEC in terms of both trade creation and trade diversion[1], and several surveys cover their main methodological approaches and shortcomings [eg. Williamson-Bottrill (1971); Verdoorn-Bochove (1972); Sellekaerst (1973); Balassa (1975) and Mayes (1978)]. The main findings are: (i) - There was trade creation which exceeded any estimates of trade diversion[2]. (ii) - No precise order of magnitude can be established since within the same year one result can be four times larger than the other. (iii) - There is a tendency for the effects to be greater if a longer time horizon is considered. So, in spite of a step-by-step process of tariff dismantling and of possible bias due to inflation and methodologies used, one can safely argue that integration has lasting effects and should not be seen as having a "once-for-all" impact. Our own estimates of weighted share indices confirm this claim.

Turning now to the approaches used we have classified them according to the "technique" employed, that is whether by: Dummy variables, elasticities or trade matrices.

The first is based on regression analysis (and is often associated with the second technique) where a dummy variable is introduced to distinguish the pre - from the post - integration periods [see: Aitken (1973) and Verdoorn-Schwartz (1972)].

The second is usually based on estimates of price elasticities with the exception of the well-known study by Balassa (1967) which used income elasticities. Some rely on hypothetical/notional values for the elasticity of substitution between domestic and foreign sources of supply, such as Verdoorn's study (1954) and other earlier studies, while most of them use estimates from import equations and a few use estimates from general equilibrium models [Prewo (1974); Miller-Spencer (1977)].

Finally, the methods based on a comparison of the trade matrices between the pre - and the post - integration periods, include the application of the RAS method and information theory [Waelbroeck (1964) and Theil (1965,1967)]; the use of gravitation models [Waelbroeck (1964)]; the share in apparent consumption [Truman (1969); EFTA (1969,1972)]; the use of the "rest of the world" share as an indication of the "anti-monde" [Williamson-Bottrill (1971); El-Agraa (1979)], and the weighted share method [Verdoorn and Schwartz (1972)].

To proceed with our own estimation of trade effects this last methodology was chosen on both theoretical and practical grounds; the latter being crucial. The reasons for rejecting the methodologies based on elasticities (and the dummy variable approach, which is a special case of this) may be stated as follows:

First, the data requirements are much larger than for the share approach. In particular, there is no need to include estimates of price changes induced by integration, which, in any case would be very unreliable given the difficulties of knowing how tariff changes and quantitative restrictions affect prices and their inevitable time-lags.

Second, the balance of payments approach developed in the previous chapter requires the provision of year-by-year data for the horizon envisaged while the estimates based on elasticities usually adopt the simplification that the effects are of a "once-for-all" type.

Third, there are the well-known difficulties of estimating the price elasticities in international trade, as recently surveyed by Stern et al (1976). These arise mainly as a result of specification difficulties (simultaneous consideration of supply and demand); aggregation bias, and "poor" price variables. In this respect it should be noted that most studies seldom go further than one digit disaggregation which also enhances the problems of using unit cost price indices and estimates of average tariff changes. But even if we were to overlook these difficulties, we would still be left with the dispute as to whether are "tariff elasticities" different from price elasticities [Krause (1962) and Kreinin (1967)].

Fourth, the particular share method chosen (the weighted share analysis) has the advantage of having a comparable measurement based on price elasticities. In fact, the Verdoorn and Schwartz (1972) study also provides a measurement based on elasticities and the results obtained point to the same magnitudes (see table 5.1 in section 5.1).

Bearing all these points in mind, it must be concluded that to reduce integration within the context of the EEC to a process of tariff changes is obviously an oversimplification; the attendant implication that price changes would in any case reflect the whole community as a concomitant of common policies is equally unsatisfactory, and we are induced to prefer residual methods, such as share analysis.

However, this is not to say that the weighted share analysis presented below is devoid of simplifications and shortcomings either in terms of assumptions concerning the implicit "anti-monde" or operational requirements (data etc.) for implementation.

4.2 - The weighted share method

There are several versions of share analysis and we have chosen the one used by Verdoorn and Schwartz (1972) which will be outlined here, together with some changes that we have introduced. We shall start by commenting on its basic features and the implicit "anti-monde". The approach has its antecedents in a paper by Major (1962) and a first presentation was given in a paper by Verdoorn and Slochtern (1964). The use of market shares is very appealing for two main reasons. First, it eliminates growth differences between import countries, and secondly the simultaneous consideration of import and export-shares takes into account the fact that the effects of tariff reductions depend not only upon the price elasticity of demand but also upon the elasticity of supply. That is, only if the elasticities of export supply are infinite will the import shares be uniquely determined by the elasticity of demand.

Its main feature, however, is that instead of using a simple arithmetic mean of the import and export shares, Verdoorn and Schwartz (1972) propose the use of a weighted harmonic mean: the weights allow for the fact that the possibilities of a country benefiting from demand expansion in a partner country are, under given supply conditions, the larger the smaller its initial share.

The basic principle in the construction of the implicit "anti-monde" is similar to that of Theil (1965) which asserts that we can predict trade flows between two regions based on the product of their total shares of world trade. This amounts to the assumption of import-export independence and it means that the exports from country i to j are large when i exports much and j imports much and that the flow from i to j becomes smaller when either i exports little and/or j imports little. Predictions are then based on the assumption that forces determining the deviation from the independence pattern are approximately constant over time.

Opposed to these simplistic views of the "anti-monde", other authors (Williamson and Bottrill 1971) have suggested that the share performance of the jth supplier in markets (rest of the world), where he neither gains nor loses preferential advantages due to integration, should be used as a control variable. Also, the introduction of home markets has been argued as essential in building an appropriate "anti-monde" through the use of shares in apparent consumption [Truman (1969), EFTA (1972) and Winters (1984)]. It is, however, possible to meet some of these limitations, as we shall show below, and another criticism by Williamson and Bottrill regarding "its apparent lack of a coherent theoretical basis" has no basis since the weighted share indices can be derived from Armington's (1969) demand system.

We shall now outline the derivation made by Verdoorn and Schwartz (1972). Excluding the home market, and the corresponding elasticity of substitution between domestic production and competing imports, they define weighted total imports (x_o) and an "average" minimal import price (p_o) as CES functions:

$$\begin{cases} x_o = \left(b_2 x_2^{-\gamma} + \ldots + b_m x_m^{-\gamma}\right)^{-\frac{1}{\gamma}} \\ p_o = \left(b_2^{\frac{1}{1+\gamma}} x_2^{\frac{\gamma}{1+\gamma}} + \ldots + b_m^{\frac{1}{1+\gamma}} x_m^{\frac{\gamma}{1+\gamma}}\right)^{\frac{1+\gamma}{\gamma}} \end{cases} \qquad 4.1)$$

where x_i (with $i=2,\ldots,m$) represents[3] the various sources of imports; b and γ are parameters.

According with cost-minimization rule we have,

$$\frac{\dfrac{\partial x}{\partial x_o}}{\dfrac{\partial x}{\partial x_i}} = \frac{p_i}{p_o} \qquad 4.2)$$

where x is total consumption

or,

$$\frac{\partial x_o}{\partial x_i} = \frac{p_i}{p_o} \qquad 4.3)$$

Differentiating the CES function we have,

$$\frac{\partial x_o}{\partial x_i} = -\frac{1}{\gamma}\left[\sum_{i=2}^{m} b_i x_i^{-\gamma}\right]^{-\frac{1}{\gamma}-1}\left(-\gamma b_i x_i^{-\gamma-1}\right) \qquad 4.4)$$

and, as $\sum_{i=2}^{m} b_i = 1$ and $\sum_{i=2}^{m} x_i = x_o$ this becomes,

$$\frac{\partial x_o}{\partial x_i} = x_o^{\gamma+1}[b_i x_i^{-(\gamma+1)}] \qquad 4.5)$$

so that the price ratio (4.3) above is:

$$\frac{p_i}{p_o} = b_i \left(\frac{x_o}{x_i} \right)^{(\gamma+1)} \qquad 4.6)$$

or, taking the reciprocal and rearranging,

$$\frac{x_o}{x_i} = \left(\frac{p_i}{p_o b_i} \right)^{\frac{1}{\gamma+1}} \qquad 4.7)$$

and,

$$x_i = x_o \left(\frac{p_i}{p_o b_i} \right)^{-\frac{1}{\gamma+1}} \qquad 4.8)$$

or

$$x_i = b_i^{-E} \; x_o \left(\frac{p_i}{p_o} \right)^{E} \qquad 4.9)$$

where $E = -\frac{1}{\gamma+1}$ is the price volume elasticity of substitution between competing imports ($E < 0$) and x_i is the quasi-share relationship for the individual importer (i) in total imports, which, when multiplied by the price ratio ($\frac{p_i}{p_o}$) gives the relationship in terms of the value shares, that is,

$$x_i p_i = b_i^{-E} \; x_o p_o \left(\frac{p_i}{p_o} \right)^{E+1} \qquad 4.10)$$

Now, to distinguish imports of a given source k from the aggregate competing sources of imports c, we have,

$$\frac{x_k p_k}{x_o p_o} = b_k^{-E} \left(\frac{p_k}{p_o} \right)^{E+1} \qquad 4.11)$$

and from the converse[4] of 4.3) we have

$$p_o = u_k p_k + (1 - u_k) p_c \qquad 4.12)$$

which, by replacing the weighted arithmetic average with a geometric one, gives

$$p_o \approx p_k^{u_k} p_c^{(1-u_k)} \qquad 4.13)$$

So that, by substitution of 4.13) in 4.11) above, we get an approximation of the share-relationship in terms of $\frac{p_k}{p_c}$:

$$\frac{x_k p_k}{x_o p_o} \approx b_k^{-E} \left(\frac{p_k}{p_c} \right)^{(1-u_k)(E+1)} \quad 4.14)$$

By moving to an index notation we can express the import-share relationship as:

$$\frac{X_{ij}}{M_j} \approx \left(\frac{p_{ij}}{p_{cj}} \right)^{(1-u_{ij})(E+1)} \quad 4.15)$$

where X_{ij} = value index

$M_j = \sum_{i=1}^{m} X_{ij}$ = total imports

u_{ij} = import share in the base year

$\frac{p_{ij}}{p_{cj}}$ = price ratio

Similarly, we can derive its logical counterpart, the export-share, as

$$\frac{X_{ij}}{B_i} \approx \left(\frac{p_{ij}}{p_{iz}} \right)^{(1-\beta_{ij})(H+1)} \quad 4.16)$$

with H > 0 and $B_i = \sum_{j=1}^{m} X_{ij}$ = total exports

However, we have that between the base year and year t the value index X_{ij} , apart from relative price changes, has also been affected by:

1 - Changes in common tariff plus outside tariffs, say a joint effect of T_{ij}^{*}.

2 - Preference shifts in demand (D_{ij}) and supply (S_{ij}).

Therefore, we can rewrite the share relationship as,

$$\begin{cases} X_{ij} \approx M_j \left(\frac{p_{ij}}{p_{cj}} \right)^{(1-u_{ij})(E+1)} D_{ij} T_{ij}^{*} \\ X_{ij} \approx B_i \left(\frac{p_{ij}}{p_{iz}} \right)^{(1-\beta_{ij})(H+1)} S_{ij} \end{cases} \quad 4.17)$$

and solving for X_{ij}, by eliminating p_{ij}, we have,

$$X_{ij} \approx M_j^{\alpha_1} B_i^{\alpha_2} T_{ij}^{*\alpha_1} \left(\frac{p_{iz}}{p_{cj}}\right)^{\alpha_3} D_{ij}^{\alpha_1} S_{ij}^{\alpha_2} \qquad 4.18)$$

where,

$$\alpha_1 = \frac{(1-\beta_{ij})(H+1)}{k}$$

$$\alpha_2 = \frac{(1-u_{ij})(E+1)}{k}$$

$$\alpha_3 = \frac{(1-\beta_{ij})(H+1)(1-u_{ij})(E+1)}{k}$$

$$k = (1-\beta_{ij})(H+1) - (1-u_{ij})(E+1)$$

In order to take better account of integration effects we should look at both export and import sides (i.e. supply and demand) and, therefore, we shall define a share index that is an appropriate index of both shares.

So, by defining a weighted share index as,

$$A_{ij} = \frac{X_{ij}}{M_j^{\alpha_1} B_i^{\alpha_2}} \qquad 4.19)$$

we have, from 4.18, that

$$A_{ij} \approx T_{ij}^{*\alpha_1} \left(\frac{p_{iz}}{p_{cj}}\right)^{\alpha_3} D_{ij}^{\alpha_1} S_{ij}^{\alpha_2} \qquad 4.20)$$

and, if we accept the standard simplification of residual imputation, by postulating that,

$$\left(\frac{p_{iz}}{p_{cj}}\right)^{\alpha_3} D_{ij}^{\alpha_1} S_{ij}^{\alpha_2} \approx 1$$

that is, that there are no preference shifts and changes in the price ratio between competing export and import markets, then

$$A_{ij} \approx T_{ij}^{*\alpha_1} \qquad 4.21)$$

So, any change in the share index (that is, $A_{ij} \neq 1$) will be attributed to tariff changes[5] and will measure the integration effects.

The specification above can be further simplified for practical work as follows:

Given that,

$$\alpha_1 = \frac{(1-\beta_{ij})(H+1)}{(1-\beta_{ij})(H+1)-(1-u_{ij})(E+1)} \qquad 4.22)$$

or,

$$\alpha_1 = \frac{(1-\beta_{ij})}{(1-\beta_{ij})+(1-u_{ij})\dfrac{(-E-1)}{(H+1)}} \tag{4.23}$$

and by assuming that there is a relationship between the price elasticities of substitution H and E such that $H \approx -E-2$ but not $H=E=-1$, we have

$$A_{ij} \approx \frac{X_{ij}}{M_j^{\frac{1-\beta_{ij}}{2-\beta_{ij}-u_{ij}}} B_i^{\frac{1-u_{ij}}{2-\beta_{ij}-u_{ij}}}} \tag{4.24}$$

or, replacing the geometric average of the denominator by an arithmetic one[6],

$$A_{ij} \approx \frac{(2-\beta_{ij}-u_{ij})X_{ij}}{(1-\beta_{ij})M_j+(1-u_{ij})B_i} \tag{4.25}$$

An important qualification to make is that A_{ij} will only be equal to one when all elements of the trade matrix are growing at the same rate. So, A_{ij} might diverge from one due to differences in the growth of trade (exports + imports) in each individual country. And, if a given country has its trade oriented to a country which experienced a faster growth in trade (not induced by integration) this will imply a value for A_{ij} different from unity (this case is identical to the devaluation situation referred to below).

Finally, we must estimate the share-index in terms of trade flows. This is done by rewriting A_{ij} in terms of current levels of trade flows. So,

$$A_{ij} \approx \frac{\dfrac{X_{ij}^t}{X_{ij}^o}}{\dfrac{1-\beta_{ij}}{2-\beta_{ij}-u_{ij}}\dfrac{M_j^t}{M_j^o}+\dfrac{1-u_{ij}}{2-\beta_{ij}-u_{ij}}\dfrac{B_i^t}{B_i^o}} \tag{4.26}$$

or,

$$A_{ij} \approx \frac{X_{ij}^t}{\dfrac{1-\beta_{ij}}{2-\beta_{ij}-u_{ij}}\dfrac{X_{ij}^o}{M_j^o}M_j^t+\dfrac{1-u_{ij}}{2-\beta_{ij}-u_{ij}}\dfrac{X_{ij}^o}{B_i^o}B_i^t} \tag{4.27}$$

Now, as $\frac{X^o_{ij}}{M^o_j}=u_{ij}$ and $\frac{X^o_{ij}}{B^o_i}=\beta_{ij}$, we can make

$$a=\frac{u_{ij}(1-\beta_{ij})}{(2-\beta_{ij}-u_{ij})} \qquad 4.28)$$

and,

$$b=\frac{\beta_{ij}(1-u_{ij})}{(2-\beta_{ij}-u_{ij})} \qquad 4.29)$$

so that 4.27) can be rewritten as,

$$A_{ij}\approx\frac{X^t_{ij}}{aM^t_j+bB^t_i} \qquad 4.30)$$

Now, let E^t_{ij} be the integration effect on flow X^t_{ij} and the "anti-monde" Y^t_{ij} and we will have,

$$Y^t_{ij}=X^t_{ij}-E^t_{ij} \qquad 4.31)$$

so that,

$$A^t_{ij}\approx\frac{Y^t_{ij}+E^t_{ij}}{aM^t_j+bB^t_i} \qquad 4.32)$$

Then, as the estimation of the "anti-monde" has to fulfill the condition[7],

$$\frac{Y^t_{ij}}{a(M^t_j-E^t_{ij})+b(B^t_i-E^t_{ij})}=A^t_{ij}=1 \qquad 4.33)$$

we can solve this system for E^t_{ij} which gives,

$$E^t_{ij}=(A^t_{ij}-1)\frac{aM^t_j+bB^t_i}{1-a-b} \qquad 4.34)$$

and this expression provides the measurement of the trade flow effects of integration.

Having summarized the weighted share technique and its main assumptions we shall now introduce corrections to reduce some of its major drawbacks.

4.3 - Extensions to the share technique

Starting with aggregation, it is easily seen that substantial bias might result from the fact that the weights used (i.e., those based on initial import and export shares) are inappropriate if calculated from aggregated data. Let us take as an example the aggregation of the imports of two countries. In this case the import share,

$$u_{ij} = \frac{X_{ij}}{\sum_{i=1}^{m} X_{ij}} \qquad 4.35)$$

would have to be equal to,

$$u_{ij} = \frac{\alpha_i X_{ij} + (1-\alpha_i) X_{ij}}{\sum_{i=1}^{m} \alpha_i X_{ij} + \sum_{i=1}^{m} (1-\alpha_i) X_{ij}} \qquad 4.36)$$

where α_i is the share of the first country and $0<\alpha_i<1$.

However, it is clear that (4.35) will only be the average of each individual import share when α_i is constant (i.e., if, for example, we aggregate the United States of America and Japan then we assume that all countries split up their exports to these two countries in exactly the same proportion), since:

$$\frac{\alpha \dfrac{X_{ij}}{\sum_{i=1}^{m} \alpha X_{ij}} + (1-\alpha) \dfrac{X_{ij}}{\sum_{i=1}^{m} (1-\alpha) X_{ij}}}{2} = \frac{X_{ij}}{\sum_{i=1}^{m} X_{ij}} \qquad 4.37)$$

So, when estimating the trade creation[8] of each individual country the sum of the effects with each partner may be quite different from the effect measured in the aggregate (the EEC). In fact, the Verdoorn and Schwartz (1972) results show an aggregation bias that ranges between -13% and + 182 % of the individual country estimates. Therefore, as a priori signs cannot be established, the trade matrices should be disaggregated as far as possible to eliminate this source of bias.

Another important source of bias in the weighted share analysis comes from what we shall call interdependence effects. That is, as the "anti-monde" is predicted on the basis of total exports and imports then the changes in mutual trade between non-participating countries will affect the estimates of trade changes between these countries and the participating ones. This can be seen from the weighted share index, defined in equation (4.25) above as:

$$A_{ij} \approx \frac{(2-\beta_{ij}-u_{ij}) X_{ij}}{(1-\beta_{ij}) M_j + (1-u_{ij}) B_i} \qquad 4.38)$$

with $(i, j = 1, 2, \ldots, m)$.

So, if we multiply a given element by a factor, say kX_{ij} (with k = 1 + % increase or decrease) then both $M_j = \sum_{i=1}^{m} X_{ij}$ and $B_i = \sum_{j=1}^{m} X_{ij}$ will be affected and all the other indices (A_{ij}) in vectors i and j will be changed so that they will show either trade diversion or creation depending on whether k is greater or less than 1, respectively. An illustration can be given through a hypothetical trade matrix where the only increase in trade was between non-participating countries, e.g.:

Year: 0

From/to	*P*1	*P*2	*NP*3	*NP*4	*TOTAL*
*P*1	0	50	70	30	150
*P*2	100	0	30	40	170
*NP*3	50	30	0	20	100
*NP*4	40	40	20	0	100
TOTAL	190	120	120	90	520

Year: t

From/to	*P*1	*P*2	*NP*3	*NP*4	*TOTAL*
*P*1	0	50	70	30	150
*P*2	100	0	30	40	170
*NP*3	50	30	0	22	102
*NP*4	40	40	22	0	102
TOTAL	190	120	122	92	524

where a 10% increase in trade between non participating[9] countries (NP) affected all elements except the ones in the upper left corner, and the weighted share indices (A_{ij}) will be:

$$\begin{matrix} 0 & 1 & .99 & .99 \\ 1 & 0 & .99 & .99 \\ .99 & .99 & 0 & 1.08 \\ .99 & .99 & 1.08 & 0 \end{matrix}$$

Furthermore, there is the problem that the larger the trade flows between non-partners, the larger the interdependence effect, although not proportional. So, if in the above matrix we multiply the non-partners' trade by a factor of 2.5 and use the same 10% increase in mutual trade, then trade diversion experienced by the participating countries will be multiplied by 1.6 and 1.86 respectively for the first and second partner and the resulting weighted share index (A_{ij}) matrix will be:

0	1	.98	.98
1	0	.98	.98
.98	.98	0	1.06
.98	.99	1.06	0

Finally, although we can know a priori the sign of the effect of changes in each individual element of the matrix, we are unable to anticipate the final outcome (specially if we already have a large degree of aggregation) in the presence of multiple changes, which can offset each other. However, we cannot assume that the interdependence effects will either be small or will offset each other. In fact, in the hypothetical matrix above they range between 1.4% and 2 % of actual flows in period t, and in the Verdoorn and Schwartz (1972) estimates the bias from interdependence in trade with the rest of the world would range between 15% and 29 % of the indices estimated (with all the indices increasing and one of them even revealing a shift from trade diversion to trade creation).

Furthermore, if, when constructing the "anti-monde", one is willing to accept that all changes in trade between participating and non-participating countries are due to integration, it nevertheless does not follow that changes in trade between non-partners are also due to integration. On the contrary, it is assumed in customs union theory that there is no trade reorientation between non-partners.

By accepting this assumption we can eliminate interdependence bias by simply rebuilding the trade matrices on the assumption that trade between non-partners remains constant[10] over the period in analysis.

It should also be noted that the interdependence effect implies that the share analysis must be conducted in quantity terms. Otherwise, any currency realignment or change in a commodity's domestic price would affect the corresponding row and column, which would show up as a trade creation/diversion effect (depending on whether it was an increase/decrease movement) while the remainder would show an opposite trade effect. Moreover, in the presence of aggregation a change in a specific commodity price will affect only a few specific bilateral flows, sometimes only one element of the trade matrix, in which case the effects would be quite different. For example, in the hypothetical matrix taken above a 10 % change in prices in the bilateral trade between non-partners would generate a total trade creation effect of 4.0 and trade diversion of 5.8 while a devaluation of the fourth country currency by 10 % would generate a total trade creation of 7.5 and total diversion of 13.5. In the absence of a trade matrix in quantities, the least we can have is a conversion to constant prices in the value matrix, although, with aggregation of commodities, this presents its own problems.

Finally, another important drawback of the Verdoorn and Schwartz (1972) approach is the exclusion of the home market in the construction of the "anti-monde", which implies that the home market is zero for all countries. In fact, recent research [Winters

(1984)] has shown that import (and concomitantly export) behaviour in terms of source allocation does not conform to the hypothesis of separability and homotheticity; that is, import allocation is not independent of domestic prices, nor are the shares invariant with respect to expenditure. However, as the weighted share index is derived from Armington's (1969) trade allocation model, which imposes both homotheticity and separability, the index necessarily excludes the home market.

Nevertheless, this can partly be solved by expanding the trade matrix through the introduction of dummy countries which correspond to the home market of each country in the matrix. This was effected for the hypothetical matrix used above by considering that the home market would be 50 units for each country, and the resulting matrix is shown below,

0	50	0	0	0	0	0	0	50
50	0	0	50	0	70	0	30	200
0	0	0	50	0	0	0	0	50
0	100	50	0	0	30	0	40	220
0	0	0	0	0	50	0	0	50
0	50	0	30	50	0	0	20	150
0	0	0	0	0	0	0	50	50
0	40	0	40	0	20	50	0	150
50	240	50	170	50	170	50	140	-

with the dotted matrices representing trade between each country and its own dummy country (home market). Thus, in lines i + 1 (i = 1,2,...,m) we have total production by destination (including the home market) and in columns j + 1 (j = 1,2,...,m) we have total consumption by supplier, and the resulting matrix of weighted share indices (A_{ij}), corresponding to a 10 % increase in intra-partners trade, is now:

0	1.07	.99	.99
1.05	0	.98	.98
.98	.99	0	0
.98	.99	0	0

On the other hand, the corresponding index matrix to the same change in the initial trade matrix (that is, assuming no home market or that it is equal to zero) is:

0	1.06	.99	.99
1.04	0	.97	.98
.98	.98	0	0
.98	.98	0	0

which corresponds to differences in total import effects ranging from 11.6% to 34.4 %. It is interesting to note that in this particular case only trade diversion is affected by the introduction of the home market.

The overall implications of introducing the home market can be analysed by using the measure of integration flow effects (E^t_{ij}) of expression 4.34) above. By substituting 4.30) into 4.34) we have,

$$E^t_{ij}=\frac{X^t_{ij}-(aM^t_j+bB^t_i)}{1-a-b} \qquad 4.39)$$

Each element (ij) of the trade matrix will be affected by home markets K_i and K_j and therefore E^t_{ij}, will become:

$$E^t_{ij}=\frac{X_{ij}+\Delta_{ij}-\left[a\left(M_j+\Delta_{ij}+\sum_{s=1,s\neq i}^{m}\Delta_{sj}+K_j\right)+b\left(B_i+\Delta_{ij}+\sum_{s=1,s\neq j}^{m}\Delta_{is}+K_i\right)\right]}{1-a-b} \qquad 4.40)$$

where Δ is the increment in X from year 0 to year t

or, by substitution of $M_j=\sum_{i=1}^{m}X_{ij}$ and $B_i=\sum_{j=1}^{m}X_{ij}$, and rearranging,

$$E^t_{ij}=\frac{X_{ij}-a(\sum_{i=1}^{m}X_{ij}+K_j)-b(\sum_{j=1}^{m}X_{ij}+K_i)}{1-a-b}+$$

$$+\frac{\Delta_{ij}(1-a-b)}{1-a-b}-\frac{a}{1-a-b}\sum_{s=1,s\neq i}^{m}\Delta_{sj}-\frac{b}{1-a-b}\sum_{s=1,s\neq j}^{m}\Delta_{is} \qquad 4.41)$$

and as the first term in the R.H.S. is zero (this result can be obtained by substituting of a and b for expressions 4.28 and 4.29 above), then the flow effects will be:

$$E^t_{ij}=\Delta_{ij}-\frac{a}{1-a-b}\sum_{s=1,s\neq i}^{m}\Delta_{sj}-\frac{b}{1-a-b}\sum_{s=1,s\neq j}^{m}\Delta_{is} \qquad 4.42)$$

It is now clear that when (as in the case above) both the jth column and the ith row remain unchanged, that is, $\sum_{s=1,s\neq i}^{m}\Delta_{sj}=0$ and $\sum_{s=1,s\neq j}^{m}\Delta_{is}=0$, then the trade creation/diversion will be unaffected by the introduction of the home market and will be Δ_{ij}. However, the corresponding diversion/creation in the unchanged elements will be affected by the home market. In this case the E_{ik} (with $k\neq j$) element will be equal to $-\frac{b_{ik}}{1-a_{ik}-b_{ik}}\Delta_{ij}$ and the E_{kj} (with $k\neq i$) element will be $-\frac{a_{kj}}{1-a_{kj}-b_{kj}}\Delta_{ij}$.

Nevertheless, the normal situation in international trade is one in which we have simultaneously $\Delta_{ij} \neq 0$, $\sum_{s=1, s\neq i}^{m} \Delta_{sj} \neq 0$ and $\sum_{s=1, s\neq j}^{m} \Delta_{is} \neq 0$ so that the effect of introducing the home market will depend on the values of $\frac{a}{1-a-b}$ and $\frac{b}{1-a-b}$ with and without the home market. As both $\frac{a}{1-a-b}$ and $\frac{b}{1-a-b}$ become smaller[11] with the introduction of the home market we may say (in general and regardless of the size of each individual home market) that for given normal positive values of Δ_{sj} and Δ_{is} the effect will be an increase of trade creation and/or decrease of trade diversion.

However, as well as considering the size of the home market, we also have to examine how it changes through time and how this would affect the estimation of integration flows. This is easily done if one splits K_i and K_j in expression (4.41) above into $k_i + \Delta_{k_i}$ and $k_j + \Delta_{k_j}$ so that,

$$E_{ij}^{t} = \Delta_{ij} - \frac{a}{1-a-b} \sum_{s=1, s\neq i}^{m} \Delta_{sj} - \frac{b}{1-a-b} \sum_{s=1, s\neq j}^{m} \Delta_{is} - \\ - \frac{a}{1-a-b} \Delta_{k_j} - \frac{b}{1-a-b} \Delta_{k_i} \qquad 4.43)$$

This alone does not show clearly how increases in both home markets (ie, $\Delta_{k_j} > 0$ and $\Delta_{k_i} > 0$) will affect trade creation and trade diversion. These will depend to a certain extent on the change experienced by the home market (see footnote 1 of section 5.1) but, if we assume that this is small, then as a result of omitting the home market there will be an underestimation of trade creation and an overestimation of trade diversion.

Note that expression 4.43) above can equally be used to measure effects of changes in home market prices and therefore the same conclusion applies. However, in the particular case dealt above (a change in one element) the result might be reversed. Take, for example, a 10 % change in the first element of between-partners trade; the resulting matrix of indices (A_{ij}) will be:

0	1.06	.98	.98
1.04	0	.98	.98
.97	.99	0	0
.97	.99	0	0

which now corresponds to differences in total import effects, relatively to the previous (no home market) situation, ranging from 22.5% to 62.6 % with a reversal of the trend. In other words, this time trade creation ($e1 = \Delta_{ij} - \frac{b}{1-a-b} \Delta_{k_i}$) is reduced and trade diversion ($e1 = -\frac{b_{ik}}{1-a_{ik}-b_{ik}} \Delta_{ij} - \frac{b_{ik}}{1-a_{ik}-b_{ik}} \Delta_{k_i}$) is increased.

Note that as these changes (either in the size or in the prices of home markets) are bound to introduce substantial effects; it then becomes crucial to know whether changes in home markets should be attributed to integration. It is useful to remember that under standard Customs Union analysis the relative size of domestic markets to home producers will be reduced either by substitution for imports or reorientation of production to export markets. Nevertheless, it is one of the basic postulates of Customs Union theory that resources released will be employed elsewhere in the economy. Furthermore, if we accept that the subsequent welfare gain can be translated into growth of income and that there are dynamic gains in terms of productivity increases due to a better allocation of resources and larger markets, then important feedback effects of integration on home markets should be expected.

However, it is essential to know whether they just offset (or, alternatively, bypass) the possible reductions so that any measured increases in the home market (that is, total production minus exports), in both cases, are entirely due to other sources of growth in the global economy. This is a question that is almost impossible to answer empirically. We will therefore choose the easiest solution by assuming[12] that they just offset so that the "anti-monde" will be constructed on the assumption that the size of domestic markets is not affected by integration.

Finally, let us turn to the question of the time pattern of the flow effects of integration. It was seen from expression 4.25) above that a condition in which the flow effects were nil was that every element of the trade matrix was growing at the same rate. An "anti-monde" based on this condition is certainly very unrealistic but it is possible to replace it by a more realistic one. As we demonstrate in the Appendix to this chapter, there is a situation in which the estimated change in the trade flows would be constant over time. This happens when, regardless of the growth rates, the absolute difference between growth rates of any two elements of the trade matrix remains constant. For example, suppose that trade between EEC, EFTA and ROW were to grow relative to the base year at 5%, 3% and 2 % in period 1, and at 7%, 5% and 4 % in period 2, and at 6%, 4% and 3% in period t, then the estimated trade flows (creation/diversion) would remain constant over that period. Just as trade was already expanding before integration, so it is certainly more realistic to assume this "anti-monde" where relative growth rates remain constant and to deduct the effect from the estimated trade flows.

We must now choose a pre-integration period that might be representative of the relative growth rates between any two elements. Having done so we must then proceed to take an average of those years that are most likely to eliminate hazardous influences. There is no a priori definition of the time horizon to consider in the estimation of integration induced trade flows. It is a matter of plotting the estimated trade flows against the calendar of tariff dismantling and then close the horizon when patterns suggest that the effects have been exhausted.

Having given a clearer presentation of the Verdoorn and Schwartz's (1972) weighted share approach and improved on their simplistic "anti-monde", we can summarise as follows:

Firstly - the weighted share technique is very sensitive to the level of aggregation both in terms of countries and commodities. Therefore, studies should be done at the highest possible level of disaggregation.

Secondly - given that the interdependence effects resulting from changes in trade between non-participating countries affect the estimation of integration induced effects, specific assumptions should be made concerning the "anti-monde" for non-participating countries. We suggested that they were assumed to keep a constant growth rate throughout the period. Furthermore, studies should be conducted in volume rather than on value terms.

Thirdly - it was shown that it is crucial to introduce the home market and that this can be easily done by introducing a dummy country into the trade matrices.

Finally - we could also prove that a more realistic "anti-monde" can be considered, whereby countries are assumed to keep their relative growth rate differentials constant, replacing the previous assumption that every country would have had the same growth rate.

In the next chapter we will show the suitability of using this improved version of the share technique to estimate the trade effects of integration.

APPENDIX - Proof that trade effects remain constant in an "anti-monde" with constant relative growth rate differentials

From expression 4.43) we have,

$$E_{ij}^{t}=\Delta_{ij}^{t}-\frac{a}{1-a-b}\sum_{s=1,s\neq i}^{m}\Delta_{sj}^{t}-\frac{b}{1-a-b}\sum_{s=1,s\neq j}^{m}\Delta_{is}^{t}-\frac{a}{1-a-b}\Delta_{k_j}^{t}-\frac{b}{1-a-b}\Delta_{k_i}^{t} \quad 1)$$

we also have,

$$\Delta_{ij}^{t}=\alpha^{o}(1+r_{\alpha}^{t})X_{ij}$$

$$\Delta_{sj}^{t}=\beta_{sj}^{o}(1+r_{\beta_{sj}}^{t})X_{sj}$$

$$\Delta_{is}^{t}=\beta_{is}^{o}(1+r_{\beta_{is}}^{t})X_{is}$$

$$\Delta_{k_j}^{t}=\gamma_{j}^{o}(1+r_{\gamma_j}^{t})K_{j}$$

$$\Delta_{k_i}^{t}=\gamma_{i}^{o}(1+r_{\gamma_i}^{t})K_{i}$$

where r^t is the percentage increase of the year zero percentage change.

Now, if we define that differences in growth rates are constant, then

$$\alpha^{o}-\beta_{sj}^{o}=c \quad 2)$$

and

$$\alpha^{o}(1+r_{\alpha}^{t})-\beta_{sj}^{o}(1+r_{\beta_{sj}}^{t})=c \quad 3)$$

so that,

$$r_{\beta_{sj}}^{t}=\frac{\alpha^{o}}{\beta_{sj}^{o}}r_{\alpha}^{t} \quad 4)$$

and similarly for all the other percentage changes, so that,

$$r_{\beta_{is}}^{t}=\frac{\alpha^{o}}{\beta_{is}^{o}}r_{\alpha}^{t}\ ;$$

$$r_{\gamma_j}^{t}=\frac{\alpha^{o}}{\gamma_{j}^{o}}r_{\alpha}^{t}\ ;$$

$$r_{\gamma_i}^{t}=\frac{\alpha^{o}}{\gamma_{i}^{o}}r_{\alpha}^{t}$$

By substitution in 1) above we have,

$$E_{ij}^t = \alpha^o X_{ij} + \alpha^o r_\alpha^t X_{ij} - \frac{a}{1-a-b} \sum_{s=1, s \neq i}^{m} (\beta_{sj}^o X_{sj} + \alpha^o r_\alpha^t X_{sj}) - \qquad 5)$$

$$\frac{b}{1-a-b} \sum_{s=1, s \neq j}^{m} (\beta_{is}^o X_{is} + \alpha^o r_\alpha^t X_{is}) - \frac{a}{1-a-b}(\gamma_j^o k_j + \alpha^o r_\alpha^t k_j) - \frac{b}{1-a-b}(\gamma_i^o k_i + \alpha^o r_\alpha^t k_i)$$

and rearranging,

$$E_{ij}^t = \alpha^o X_{ij}$$

$$-\frac{a}{1-a-b} \sum_{s=1, s \neq i}^{m} \beta_{sj}^o X_{sj} - \frac{b}{1-a-b} \sum_{s=1, s \neq j}^{m} \beta_{is}^o X_{is} - \frac{a}{1-a-b} \gamma_j^o k_j - \frac{b}{1-a-b} \gamma_i^o k_i +$$

$$+\alpha^o r_\alpha^t \left[X_{ij} - \frac{a}{1-a-b} \sum_{s=1, s \neq i}^{m} X_{sj} - \frac{b}{1-a-b} \sum_{s=1, s \neq j}^{m} X_{is} - \frac{a}{1-a-b} k_j - \frac{b}{1-a-b} k_i \right] \qquad 6)$$

Now as we have that,

$$\sum_{s=1, s \neq i}^{m} X_{sj} = \sum_{i=1}^{m} X_{ij} + k_j - X_{ij} - k_j \qquad 7)$$

and

$$\sum_{s=1, s \neq j}^{m} X_{is} = \sum_{j=1}^{m} X_{ij} + k_i - X_{ij} - k_i \qquad 8)$$

the term in [...] brackets becomes,

$$[\ldots] = \frac{X_{ij} - a\left(\sum_{i=1}^{m} X_{ij} + k_j\right) - b\left(\sum_{j=1}^{m} X_{ij} + k_i\right)}{1-a-b} \qquad 9)$$

which we know from 4.41) to be zero.

Then, the integration effects on year t, are:

$$E_{ij}^t = \Delta_{ij}^o - \frac{a}{1-a-b} \sum_{s=1, s \neq i}^{m} \Delta_{sj}^o - \frac{b}{1-a-b} \sum_{s=1, s \neq j}^{m} \Delta_{is}^o - \qquad 10)$$

$$-\frac{a}{1-a-b} \Delta_{k_j}^o - \frac{b}{1-a-b} \Delta_{k_i}^o$$

That is, the integration effects will be equal to those of the base year.

1. See in section 2.2 why these are not necessarily equivalent to the trade flows estimated.

2. The only exception is the study by Resnick-Truman (1974) whose results are analysed in Balassa (1975).

3. Note that the procedure used below of introducing a home market dummy country (say $b_1 x_1^{-\gamma}$) is not incompatible with the share analysis. Therefore, x_1 is left to represent domestic sources of supply.

4. As, $p_o = \psi(p_k, p_c)$ which by total differentiation gives: $dp_o = \frac{\partial \psi}{\partial p_k} dp_k + \frac{\partial \psi}{\partial p_c} dp_c$ then it must be of the form $\frac{x_k}{x_o} dp_k + \frac{x_c}{x_o} dp_c$ or $\frac{dp_o}{p_o} = \frac{x_k p_k}{x_o p_o} \frac{dp_k}{p_k} + \frac{x_c p_c}{x_o p_o} \frac{dp_c}{p_c}$ and with $u_k = \frac{x_k p_k}{x_o p_o}$ we have that $\frac{dp_o}{p_o} = u_k (\frac{dp_k}{p_k}) + (1 - u_k)(\frac{dp_c}{p_c})$

5. The implicit tariff change can then be estimated as $T_{ij}^* = A_{ij}^{\frac{1}{\alpha_1}}$

6. Note that A_{ij} is in fact a weighted harmonic mean of import and export shares. As $\left[\frac{1}{\frac{1}{(\frac{X_{ij}}{M_j})} + \frac{1}{(\frac{X_{ij}}{B_i})}} \right]$ gives $\frac{X_{ij}}{M_j + B_i}$

7. It must be noted that in a multi-country case M_j^f and B_i^f should also be corrected for integration effects on other countries, that is, $E_{ij}^f + \sum_{s=1, s \neq i}^{m} E_{sj}$ and $E_{ij}^f + \sum_{s=1, s \neq j}^{m} E_{is}$, instead of only E_{ij}^f. Therefore, E_{ij}^f estimates will always be biased except in the case where these two summations are zero, otherwise, it will be biased downwards whenever they are greater than zero (and vice versa) and will be dependent on the relative magnitude when one is greater than zero and the other one is smaller than zero.

8. We shall now start using creation and diversion terms while the correct words would be increased or decreased trade flows. This is done for reasons of convenience, for as has already been stated in chapter II the terms trade creation and trade diversion do not exhaust all integration effects.

9. Note that even if we assume no aggregation bias and use just one "rest of the world" country then even the intra-partners trade estimates would be affected.

10. Or rather that relative growth rates remain constant. An alternative could be the assumption of no trade between partners but this would introduce an upward bias in the estimates of trade effects.

11. This results from a much larger increase in the denominator as can be observed in the corresponding expressions:

$$\frac{a}{1-a-b}=$$

$$\frac{X_{ij}K_i+X_{ij}\left(\sum_{j=1}^{m}X_{ij}-X_{ij}\right)-\frac{X_{ij}}{2(\sum_{i=1}^{m}X_{ij}+K_j)}-\frac{X_{ij}}{2(\sum_{j=1}^{m}X_{ij}+K_i)}}{\left[K_j(\sum_{j=1}^{m}X_{ij}-X_{ij})+K_i(\sum_{i=1}^{m}X_{ij}-X_{ij})+K_iK_j\right]+\left(\sum_{j=1}^{m}X_{ij}\sum_{i=1}^{m}X_{ij}-X_{ij}\sum_{i=1}^{m}X_{ij}-X_{ij}\sum_{j=1}^{m}X_{ij}-X_{ij}^2\right)}$$

and similarly for $\frac{b}{1-a-b}$.

12. As it will be shown in subsequent chapters this assumption will make our results overoptimistic.

Chapter Five

MEASURING EEC INTEGRATION EFFECTS

5.1 - Effects on trade flows of industrial products

Some qualifications regarding the data and their limitations need to be mentioned before presenting results obtained through the approach outlined in the previous chapter. Data sources are the OECD Foreign Trade Statistics (series B and C) and EUROSTAT. Yearly trade matrices were constructed and the usual caveat on trade matrices due to discontinuities and differences in presentation of data (CIF/FOB) applies. In the present case, for reasons of availability, we use import data and, since disaggregation is not complete, a "ROW" (rest of the world) country was constructed so that its imports are the difference between total exports (in FOB values) and individual exports (generally in CIF values).

Nevertheless, it should be noted that although disaggregation at the country level has gone beyond the usual practice in integration studies, the level of disaggregation at the commodity level is still not fully satisfactory. For the period 1958-72, total trade was used which precludes any separation of effects (other than budgetary) due to the Common Agricultural Policy, the effects of which are present from at least 1968. On the other hand, from 1972 on, disaggregation is used at the one digit level, which, although not entirely satisfactory, at least allows for the exclusion of oil trade, which could introduce a substantial bias (especially in the U.K. case), and also allows for the introduction of Common Agricultural Policy (CAP) estimates. Owing to classification difficulties, the introduction of the home market was only possible for the period 1961-72 using National Accounts Data (EUROSTAT) of GDP, and for the period 1973-81 for commodities of SITC groups 0+1, 2, 5 and 6.

It should also be stressed that the trade matrices used are value matrices and they were only converted to constant (1975) prices for the 1958-72 period, using each country's average import price. As this is not equal to each element price index it only works as a normalization procedure and therefore some price effects will still be present.

Finally, the base years used were 1958-60 for the period 1961-72, and 1972-74 for the period 1973-81. The use of a three-year average is intended to allow for the fact that the year before integration might be abnormal and that there are lags/leads in the reactions of economic agents to integration. For similar reasons the post-integration annual values were also taken as a three-year average so that circumstantial fluctuations are not picked up as integration effects.

Given these preliminary qualifications, the results for the ***period 1961-72*** are presented below. We shall begin by comparing our estimates with those of previous studies presented in table 5.1 below.

Table 5.1 - Comparison of Estimates of Trade Creation and Diversion

(million US dollars)

	Trade Creation (1)					Trade Diversion (2)			
	R-T 1968	V-S (3) 1969	V-S (4) 1969	Own (5) 1969	Own (6) 1969	R-T 1968	V-S (3) 1969	V-S (4) 1969	Own (5) 1969
Germany	-659	4141	4272	6654	1096	1732	267	654	2518
France	582	3321	2937	1576	1140	737	248	311	5562
Italy	1022	1490	1543	3770	477	62	154	214	329
Netherl.	93	1084	651	3946	437	190	216	130	543
Belg-Lux	152	1096	1033	3020	367	281	183	325	651
TOTAL	1190	11132	10436	18966	3517	3002	1068	1634	9603

Sources: Mayes (1978), Verdoorn and Schwartz (1972) and our own estimates

Notes:

1) Refers to: trade creation + external trade creation + trade reorientation - trade suppression (i.e. total imports in my study).

2) Includes: external trade creation - trade diversion - trade suppression (i.e. total imports from the rest of the world).

3) Estimation using an import equation with a dummy for integration.

4) Own estimate using V-S data and partial correction for interdependence bias (i.e. trade between non-partners is held constant).

5) Our own data (i.e. total trade instead of manufactured goods) and converted to 1969 prices using the implicit import price deflator. Includes the home market of EEC5 countries plus the one of the U.K.

6) External trade creation, i.e.: positive changes in trade with non-partners. It is also included in (2).

The main conclusion to be drawn is that our results not only confirm the previous generally accepted evidence that trade creation is likely to exceed any diversion, but also point to the fact that previous estimates have been heavily downward biased. However, this is largely overestimated because we only had data for the home market for member countries which are expected to experience trade creation. If we had the data for the home market for non-participating countries where trade diversion is expected, but might be underestimated, then the total trade effect would be smaller.

A large differential might still be expected even if the bias in our own estimates could be eliminated. Note from previous chapter (equations 4.42 and 4.43) that integration effects are smaller[1] when home markets are excluded. Nevertheless, our results are also affected by the fact that some elements of the trade matrix were assumed as zero, results in biased estimates for the "rest of the world country", resulting from aggregation and omission of intra-"rest of the world" trade. On balance, if these offset the previous one the final outcome will be close to the actual value.

Figure 5.1

WEIGHTED SHARE INDICES FOR GERMANY - 1961-80 - TOTAL TRADE

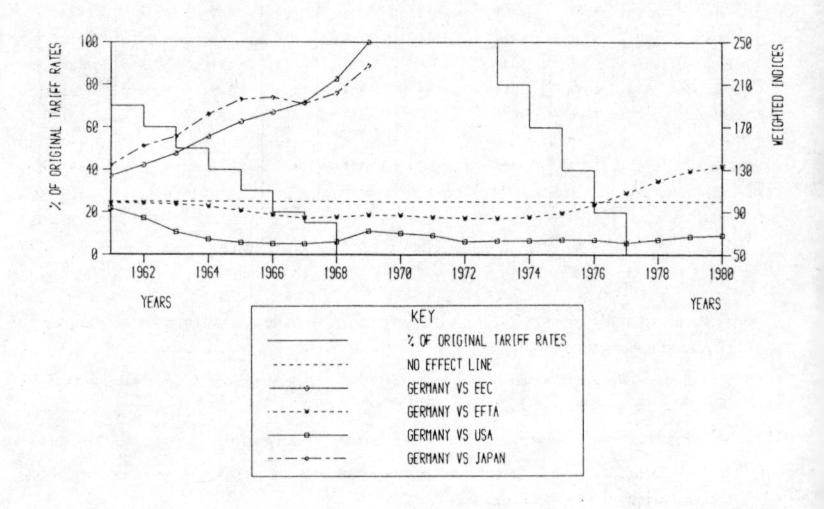

Another important result is that the amount of trade diverted is rather small, except in the case of France. The estimate for France cannot, however, be entirely ascribed to integration. It probably results from the fact that we are using total trade which (contrary to manufacturing) includes the changes in trade with its former colonies which are also included in the rest of the world country. In fact, this represents around 98 % of total trade diversion and an observation of country-by-country values shows that it only experienced a small amount of trade diversion in relation to a few EFTA countries.

Table 5.2 - Total trade effects of integration (1961-72)

million ECU's at 1975 prices

countries	Germany		France		Italy		Netherlands		Belgium-Lux.	
Years	X	M	X	M	X	M	X	M	X	M
1961	657	948	172	522	532	1298	272	472	270	323
1962	1251	1748	97	994	937	2581	545	956	585	583
1963	1524	2365	76	1432	1362	3175	924	1470	998	951
1964	2167	3168	269	1551	1963	3383	1309	1981	1472	1336
1965	2947	3734	339	1607	2641	3356	1657	2434	1857	1778
1966	4113	3860	470	1577	3367	4172	2006	2703	2163	1987
1967	5440	3891	514	1705	4143	4875	2491	3212	2554	2373
1968	6987	4873	698	2029	4998	5917	3295	4007	3328	2987
1969	8614	7252	1284	2154	5742	7238	4296	5216	4260	3865
1970	10854	10289	1950	2248	6270	8371	5473	6300	4890	4567
1971	12504	12476	3163	2731	7110	9046	6764	7208	5772	5245
1972	15539	14417	4500	4189	7551	9049	8386	8478	7026	6542

X = Exports; M = Imports

In contrast we find that the total amount of external trade creation was rather large and in the case of Italy it was even greater than trade diversion. The countries which benefited from this external trade creation were the U.K., Ireland, Greece, Yugoslavia and Japan. The United States was only slightly affected by trade diversion because the trade diverted by Germany, Italy and Belgium was partly offset by external trade creation with France and the Netherlands.

If we match the phasing of tariff dismantling with the time pattern of integration effects, (e.g. in Fig. 5.1 above for Germany), we can see that they conform to the hypothesis that the effects are long lasting and therefore consistent with the balance of payments framework previously developed.

It should also be noted that the choice of 1972 as the closing year, although convenient because of the enlargement of the Community after that year and association with EFTA, is also consistent with the time pattern presented by the integration effects.

Finally, the trade effects on the balance of payments of each individual country can be seen in table 5.2, above.

Individual countries present some differences: German values are the largest in absolute terms, while Italy has the largest relative effects. It is also interesting to note that after an initial worsening (say a shock effect) of the trade balance all of them present a recovery although Italy and Netherlands are still left with a negative effect on the balance of payments.

Having examined the trade effects for the first period let us proceed by presenting the estimated values for the enlarged community in the ***period 1974-81***. From table 5.3 presented below it can be seen that, as one would expect, the effects for the former five countries are much smaller than in the previous period (with the exception of France for the reasons mentioned above). However, what is a very important feature, and what distinguishes it from the previous period, is that (with the exception of Belgium-Luxembourg) all the older members start by improving their trade balance at the expense of the new members; this is seen most clearly in the penetration of the U.K. market. It is also important to note that during this period a large amount of trade creation was experienced by all countries (not just the U.K.) but,

Table 5.3 - Total trade effects of integration for manufactured goods (1974-81)

million ECU's at 1975 prices

years		1974	1975	1976	1977	1978	1979	1980	1981
Germany	X	1850	3215	2013	2544	1764	4180	3902(2)	2727(2)
	M	1558	3618	6057	8286	10515	13604	10831	7138
France	X	1883	3414	3553	3702	4188	5281	4916	3789
	M	72	1556	2489	3425	3982	5172	5438	4997
Italy	X	782	1543	1317	1656	2376	2789	1977	971
	M	931	1453	1020	1585	2653	3927	3827	2822
Netherl.	X	914	1801	712(1)	355(1)	754(1)	2578	6773(2)	7926(2)
	M	1159	2259	5129	3916	2675	4783	4134	3001
Belg-Lux	X	257	660	869	985	815	1935	1224	645
	M	296	1101	1754	2328	2593	3762	2593	1687
U.K.	X	202	416	-353	144	826	2672	1727	757
	M	1789	1470	1031	1798	3415	4744	3944	2787
Ireland	X	55	133	201	284	392	486	635	678
	M	73	123	316	471	774	804	804	717
Denmark	X	101	116	37	-29	-41	33	74	49
	M	172	393	507	579	432	259	134(2)	26(2)

Notes:

(1) - Netherlands estimates were corrected for an upward bias in the measurement of trade in manufactured goods with the "rest of the world".

(2) - Corrected for changes in the transport industry.

contrary to the previous period, trade diversion was also very large, in particular with respect to trade with the "rest of the world" and, to a smaller extent, with EFTA (with Germany and France as exceptions). In contrast there was an increase in trade with the United States (with the exception of Italy, Germany and Denmark).

Looking now at the new members' performance (U.K., Ireland and Denmark) it was observed that they experienced larger trade diversion than external trade creation and in a few cases the net effect almost offsets the trade creation with EEC (with the exception of Ireland's performance with regard to most industrial products[2]). Not surprisingly all of them have shown trade diversion in relation to EFTA, to which two of them belonged before joining the EEC, while the former EEC members present a small amount of trade creation. The case of countries moving from a free trade area to a customs union represents a situation seldom dealt with in customs union theory and, unless we are prepared to accept that enlargement had a downward effect on prices which left EFTA producers out of the market, we will have to question the existence of a unique "world price" or resort to some preference shift hypothesis. In contrast with this result we find (Mendes 1985a) that in trade with the United States all countries but Italy, Germany and Denmark experienced a small amount of trade creation.

It is also important to note that there are significant differences at the commodity level. To begin with only food and beverages revealed a generalised and strong diversion effect (vide next section on CAP). Also, trade in raw material and intermediate goods (classes 2,4 and 5) showed large fluctuations and although the new members showed a trend for trade diversion, countries, in general, improved their trade balance. However, the performance was not the same in the most important sector of manufactured goods (class 6) and Germany, Italy, U.K. and Denmark experienced trade diversion in relation to the USA.

Given that some of the sources of bias offset each other, and that results are consistent with a priori expectations, we conclude that the estimates are not substantially different from the actual values, and we shall use them later in applying the balance of payments framework.

5.2 - The balance of payments effects of the CAP

The effects of the Common Agricultural Policy (CAP) have generally been excluded from the studies of the effects of European Integration because it requires separate treatment. In fact, although it also uses tariffs (a common external tariff on a few agricultural commodities - coffee, tea, tobacco and fish), its main instruments are different.

CAP has replaced national intervention schemes in agriculture by implementing a Common Market for most temperate zone food products (see list in table 5.4 below). The objectives of the policy are the improvements in the living standards of the farmers, and self-

Table 5.4 - Main Markets and CAP Intervention Schemes

Policies		Planning				Intervention					External Protection						
Markets/food products (% of total FEOGA guarantee payments)		Target price	Guide price	Norm price	Basic price	Intervention price	Withdrawal price	minimum price	production aid	deficiency payment	threshold price	Sluice gate price	reference price	Levies	Supplementary levies	Tariffs	Export restitutions
cereals	Common wheat	X				X			X(a)		X			X			X
and	Durum wheat	X				X					X			X			X
rice	Barley	X				X					X			X			X
	Rye	X				X					X			X			X
(13.3%)	Maize	X				X					X			X			X
	Rice	X				X					X			X			X
Sugar	white sugar	X				X					X			X			X
(7.1%)	Sugar beet							X									
oils	oilseeds of																
and	sunflower	X				X				X							
fats	soya bean		X							X							
	cotton								X								
(5.7%)	olive oil	X															
milk	butter					X					X			X			X
products	skimmed dryed milk					X					X			X			X
(38.8%)	cheese (c)					X					X			X			X
beef,veal,	live cows		X														
pigmeat	carcase beef					X							X		X	X	
(10.4%)	pigmeat				X	X						X			X		X
4.1%	Fruit and Vegetables				X		X						X			X	X
2.1%	Wine		X			X(b)							X			X	X
3.2%	Tobacco			X													
Other	peas and beans	X(d)						X		X							
	Eggs											X			X		X
(4.3%)	Poultry											X			X		X
	Fish		X			X(e)	X						X			X	X

Sources - Berends (1983) and Official Journal 31.12.82

Notes:a) only to specified regions; b) to storage and distillation; c) only in Italy; d) activity price; e) only sardines and anchovy sufficiency of food, through the implementation of three basic principles: (1) - free trade with the same price level within the EEC; (2) - preferential pricing, and (3) - common financing of the intervention scheme. In practical terms this has meant the creation of a highly protected market for the benefit of farmers and countries with an interest in exporting food commodities, which has been accomplished by maintaining a minimum import price for food from the rest of the world and providing export subsidies enabling production in excess of supply within the EEC to be sold in international markets. It has also provided some buy-in intervention arrangements and subsidies to producers.

CAP has been implemented by the European Agricultural Guidance and Guarantee Fund (known as FEOGA, the French acronym). This has a guidance section aimed at the provision of funds for the structural improvement of the agricultural sector. However, its role is rather limited, accounting for only around 4% of the CAP spending, most of which went in co-financing of projects for land improvement, irrigation and networks of milk distribution (36%), and has been largely concentrated in three countries: Italy, Germany and France (75% in the period 1964-76). Therefore, there will be no significant price effects to consider and in the case of the Guidance section we can confine ourselves to budgetary costs alone.

On the other hand, the Guarantee section of the FEOGA has a significant impact on the balance of payments both because of the large budget flows involved and because of the price effects on trade in agricultural goods. The role of the Guarantee section is to finance the commitments of a Common Market for agricultural commodities.

To fulfil its purpose several mechanisms are being used for the different markets covered by the CAP, which are outlined in table 5.4 above in which they are grouped under the three main objective headings. Both these mechanisms and their emphasis have experienced changes during the period analysed (1973-81) but, in general, the table provides a good picture of the CAP schemes. In particular, the main markets covered have the same mechanisms, that is, an intervention price ensuring that farmers will not have unsold production; a high level of protection from external competition by means of levies, and easy access to the international markets through subsidized sales. Such a policy can be seen formally as a policy of domestic protection through tariffs (levies) and export promotion through subsidies, with the induced changes in international trade of these commodities reflected in each country's balance of payments.

However, the common financing of such a scheme through the EEC budget and the preferential price system adopted transform what could otherwise be a national cost (assuming that each country could have a similar intervention scheme supported by its own exchequer) into a cost across the exchanges which obviously has important balance of payment effects. Indeed, the balance of payments effects of CAP are being increasingly acknowledged as the most appropriate way of measuring each country's outcome from membership to the EEC [vide resumé of discussion on the CAP effects on the balance of payments in Whitby (1979)] and more recent research has concentrated on two of these effects, that is, budget costs and the increased price of food in intra-EEC trade. So, there is recognition that discussions of the redistributive effects of CAP based exclusively on net budget payments[3] are only a part of the story and that the trade effects[4] of the increased price of food, as suggested earlier by Kaldor (1971) and Miller (1971), are equally important.

Studies which have attempted to measure these two effects include Blancus (1978), Koester (1978), Rollo-Warwick (1979) and CEPG (1979). The basic approach used is illustrated in the following supply-demand diagrams:

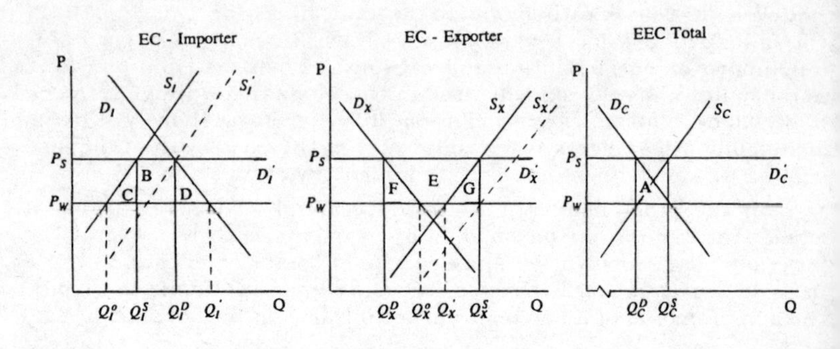

If prior to integration there were a situation of free trade, and no common budget, then in the case of the importer the position would be one of traditional introduction of protection which could be effected by either introducing a tariff so that the domestic price would be P_s or paying to farmers a subsidy of $P_s - P_w$. In the later case there would not be any consumers surplus loss (triangle D) and the direct effect on the trade balance would be only a reduction of imports by $\overline{Q_I^oQ_I^s}$, while in the case of a tariff there would be a decrease of $\overline{Q_I^oQ_I^s}+\overline{Q_I^DQ_I'}$. It is, therefore, crucial to know what the position was (or what it is assumed to have been) before integration. However, due to the large number of mechanisms both within a country and between countries, it seems that there is no way to avoid the arbitrary nature of selecting an assumption relative to the position before integration (Koester 1978). The basic assumption generally used is that each individual country would use an intervention price of the same level (P_s, as if there was CAP) and that therefore the only cost to measure across the exchanges is the rectangle B which is made up of two parts: (1) the levies paid on imports from third countries plus (2) the increase in price of food imported from partner countries. It is argued (Rollo-Warwick 1979) that the welfare losses (C+D) are not a corollary to EEC membership because each country might have chosen to support farm prices at P_s in any case. However, if it is easy to accept this hypothesis concerning the production effect (C) it is obvious that it is the CAP which is responsible for the increase in prices to consumers and the consequent welfare loss (D). In fact, the way the CAP has been operating has meant much higher producer prices than the prevailing international ones, while most countries used to keep international prices and subsidize farmers (eg. the deficiency payment system used in the U.K. prior to integration). If subsidization was more common in the pre-CAP situation, then the move to a "tariff system" should be considered a result of integration. World prices are therefore used as the appropriate reference[5] and the supply curves can be seen to shift to the left (the dotted curves S_I' and S_X') with a corresponding reduction in imports $(Q_I^D Q_I')$ and an increase in exports $(Q_X^D Q_X^o)$. In this way we can measure the effects on the volume of trade flow which cannot now be ignored. These would in any case be one of the effects of a system based on a common administrative price which is supposed to prevail throughout a large and heterogeneous market. Furthermore these effects are also the result of lobbying by different farming groups. It is therefore evident that regions or countries with a comparative advantage in both the natural factors and technical endowments necessary to produce the most favoured and surplus commodities will have substantial gains and the pattern of specialisation and international trade within the community will be strongly affected and will be reflected in each country's balance of trade.

However, the argument that elasticities in the agricultural context are meaningless is, in any case, an insufficient reason to exclude any trade flow changes and, in our view, has been widely (ab)used when evaluating the CAP. The standard argument says that there is no "true" world market for food products that can be used to provide a measure of true opportunity costs and that the world market represents only a small share of total production, because trade in foodstuffs is the subject of bilateral and government transactions. On the other hand it has also been argued that the EEC market is too large to enable the assumption of infinite elasticity of world supply, and it is commonly accepted that the CAP has induced a downward bias in "international" prices. Furthermore, there are the computational complexities arising from the use of unit cost values and the existence of significant entrepot trade which introduce a substantial bias in the estimation of world and domestic prices.

Nevertheless, although one might in principle agree with these arguments we have to acknowledge that the alternatives suggested, that is, using notional values for elasticities or excluding any volume trade effects, are quite unrealistic given the changes in the trade pattern verified after integration.

In this study we will use a trade volume effect estimated through the share technique described above. In fact, once we accept the core assumption that any shifts are due to integration, there is no strong theoretical reasons to consider agriculture as different from industrial goods since the CAP can be considered as being equivalent to the introduction of a tariff/subsidy. It has the further advantage of not excluding the shifts in trade arising from what could be called the "taste/proximity" effects derived from the quality, technical and market organisation standards introduced by CAP rules and which cannot be directly translated into price changes.

Returning to the diagrams above it can easily be seen that in the case of the exporting country the increased income from the subsidized volume of exports would be given by the area E+F+G. This is also made up of two parts - restitutions on exports to third countries and the increased price of exports to EEC members. Here again we have welfare losses, which, contrary to Rollo and Warwick's (1979) assumption, should be attributed to EEC membership. It is, indeed, difficult to assume that a national policy would bring domestic prices to consumers at P_s, and is therefore likely that $\overline{Q_X^D Q_X^o}$ would go on being consumed domestically so that the increased volume of exports would be $\overline{Q_X' Q_X^S}$.

Finally, the diagram for the EEC total represents the total budget payments made by FEOGA to subsidize exports to third countries. This is not paid entirely by levies on imports but also from VAT contributions. In fact, since the EEC budget is unified, without any specified allocation of receipts, the estimation of each country's contribution to the CAP is based on its share of total contributions to the EEC budget.

There are also some indirect effects of the CAP arising through the multiplier; the constraints on a member country's choice of policies, and the costs of balance of payments adjustments. However, since the direct effects will be introduced in the model developed to measure growth effects, which is basically dynamic, these indirect effects will be partly accounted for through the changes in the terms of trade and in the income elasticity of demand for imports.

We shall now proceed by presenting the corresponding estimates for each of the flows referred to: the budget; higher prices of food, and the trade volume effects.

(a) - The net budget effects (a negative sign means net contribution) resulting from membership of the Common Agriculture Policy presented in table 5.5 below were drawn from official accounting figures produced by the Community and are corrected for the Monetary Compensatory Amounts (MCA).

That is, we need to avoid the double counting[6] that could arise because of the "exporter pays" system under which the U.K. and Italy import MCA's are not paid to them but to the exporters. These are then deducted from the other member's receipts and added to the ones of the U.K. and Italy.

Table 5.5 - Net budget flows with the FEOGA

in million ECU's at current prices

Country	1973	1974	1975	1976	1977	1978	1979	1980	1981
Germany	-223	-209	-301	-623.9	-485.7	-544.8	-879.7	-963.4	-1081.6
France	137	95	224	225	57.6	341.6	174.3	653.1	895.3
Italy	0	74	173	-41.9	-259.3	419.8	322.7	437.6	558.2
Netherl	-56	-44	-54	220	187.8	122.2	434.2	583.4	366.5
Belg+Lux	-103	-99	-123	-21	-9.4	-34.2	50.4	-140.4	-130.0
U.Kingdom	88	25	-134	-171.1	-154.9	-285.1	-907.1	-1449.9	-1288.8
Ireland	50	49	77	167.7	375.9	300.1	409.2	502.7	396.9
Denmark	107	110	138	245.4	287.5	362.8	395.0	378.8	299.0

Source: Mendes (1985a)

(b) - The higher price of food is calculated by multiplying the imports from member countries by the corresponding differential between intervention and world prices. However, it is doubtful whether the total effect should be accounted. In fact, the specific characteristics of food trade, namely perishability, transportation requirements and marketing practices, condition trade in such a way that this is very often carried out with neighbours whatever the price conditions. It is then unrealistic to assume, in spite of higher prices within the community, that all intra-EEC trade is trade diversion. Therefore we shall adopt the assumption that at least 50% of intra-EEC trade would always exist in spite of higher prices and that world prices would be around 10% higher if it were not for the CAP. So, by multiplying 90% of the price differential by half the intra-EEC trade in food goods we obtained the estimate[7] of the net effects of increased prices which is presented in table 5.6 below.

(c) - The increased volume of total exports and imports was estimated using the share technique and the corresponding net effect on the balance of trade is presented in table 5.7 below.

Given these results we are now in a position to assess the overall direct effect of the CAP on the balance of payments of each member country. From table 5.8 below it can be seen that Germany, U.K., Belgium-Luxembourg, Italy and France were the countries whose balance of payments were negatively affected.

The only ones to benefit were the Netherlands, Ireland and Denmark. It is interesting to note that France, which is commonly regarded as the great beneficiary, does not appear to fare well on balance of payments grounds. The reason is that the negative trade effects (which in previous studies have been disregarded) more than offset the positive effects from budget transfers and increased prices of food. A similar outcome can be identified for Italy, although, here the offsetting factor is the higher cost that it has to pay for its imports

Table 5.6 - Net balance of payments effects of increased prices of food

in million ECU's at current prices

Country	1973	1974	1975	1976	1977	1978	1979	1980	1981
Germany	-51.2	4.5	12.1	18.2	272.9	251.9	396.3	661.2	1137.1
France	182.1	179.7	169.6	186.8	79.7	181.4	204.6	160.6	-114.3
Italy	-148.6	-140.1	-213.3	-260.9	-460.9	-669.9	-850.6	-1039.6	-1048.3
Netherl.	148.8	169.1	186.2	207.6	121.6	155.8	229.1	184.4	61.9
Belg-Lux	-24.7	-26.0	-34.2	-60.5	44.9	55.4	96.8	121.2	164.1
U.K.	-255.6	-322.2	-280.9	-222.4	-208.3	-141.5	-239.0	-226.3	-183.6
Ireland	57.6	56.1	71.7	65.9	75.4	65.4	56.4	62.9	14.8
Denmark	91.7	79	88.8	65.3	74.6	101.5	106.4	75.5	-31.8

Sources of data: - Trade flows between EEC partners correspond to trade matrices built from OECD statistics and covers 0+1 categories of SITC classification

- The EC index of producer prices of agricultural products was used as price index of imports from partner countries - Eurostat 1-1983.

- The world price index was estimated from NIMEXE - Eurostat Statistics of imports from Class 1 countries (other industrialised) using the unit price of commodities 01 to 25 and taking 1976 as the base year.

- The initial price differential (1976) was taken from Blancus (1976).

Note: The changes in each trade flow were obtained by multiplying the actual values by a differential factor (F) equal to: $F^t = .45\left(1 - \frac{P_w^t}{P_c^t}D\right)$ with: P_w^t = price index of world prices; P_c^t = price index of CAP prices, and D = ratio of international/CAP prices in 1976.

of food.

Another important finding to note is that in the U.K., France and Italy the negative trend in the balance of payments in the late part of the period is reversed because of a substantial increase in the amount of trade diversion which exceeded trade creation. In fact, trade diversion usually offsets external trade creation and a country by country analysis showed that in some cases it is as large as trade creation.

The results support the widespread claim that CAP was responsible for large amounts of trade diversion, confirming its role as a source of economic inefficiency. The results also make clear that (at least until 1979) even as an instrument of protection the scheme does not necessarily mean balance of payments improvements. These results will be used in the next chapter when we consider the growth effects of integration. Meanwhile let us turn to the budgetary question.

Table 5.7 - Net trade effects of CAP

in million ECU's at current prices

Country	1973	1974	1975	1976	1977	1978	1979	1980	1981
Germany	.	-11	-225	-776	-1221	-1208	-1068	-544	-152
France	.	-393	-537	-1014	-1072	-908	-7	630	1111
Italy	.	-33	93	191	183	249	163	771	1329
Netherlands	.	309	577	885	1137	1315	1544	1788	2097
Belgium+Lux	.	-37	-143	-292	-398	-424	-377	-320	-264
U.K.	.	-269	-133	13	233	272	648	2530	4289
Ireland	.	109	173	278	379	503	600	527	518
Denmark	.	188	230	235	305	448	614	72	875

Source: Mendes (1985a).

Table 5.8 - Net balance of payments effect of the CAP

in million ECU's at current prices

Country	1973	1974	1975	1976	1977	1978	1979	1980	1981
Germany	-274	-216	-514	-1382	-1434	-1501	-1551	-846	-97
France	319	-118	-143	-602	-935	-385	372	1444	1892
Italy	-149	-99	53	-112	-537	-1	-365	169	839
Netherl.	93	434	709	1313	1446	1593	2207	2556	2525
Belg-Lux	-128	-162	-300	-374	-363	-403	-230	-339	-230
U.K	-168	-566	-548	-381	-130	-155	-498	854	2817
Ireland	108	214	322	512	830	869	1066	1093	930
Denmark	199	377	457	546	667	912	1115	1183	1142

Source: Mendes (1985a).

5.3 - Net budget

Although one would expect to encounter some difficulties in measuring the receipts from the Community we found that there were also some problems concerning the contribution side[8]. The questions of how the budgetary flows affect the balance of payments and which components should be taken into account have not been settled. A major difficulty is the inclusion of the "non-budget" component of the communities, that is, the European Coal and Steel Community, the European Development Fund (EDF) and the Euratom which have different financing mechanisms, including outside financing, and which accounted for 5-7% of total receipts of the EEC in the 1970's. However, as their operations deal mostly with special

Table 5.9 - Annual net flows from the Budget 1958-1972

in million U.A. at current prices

Year	Germany	France	Italy	Netherlands	Belgium-Lux.
1958	-3.0	-2.6	-2.0	-1.0	8.6
1959	-6.9	-6	-4.5	-2.3	19.6
1960	-8.0	-6.9	-5.2	-2.6	22.5
1961	-9.5	-8.1	-6.2	-3.1	26.9
1962	-12.0	-10.9	-5.6	-4.3	32.9
1963	-12.8	-11.4	-6.9	-4.4	35.4
1964	-14.7	-12.8	-8.7	-4.9	41.0
1965	-19.1	-13.4	-9.9	-3.9	46.0
1966	-23.6	-14.5	-11.4	-3.2	52.7
1967	-50.2	-4.3	-15.1	12.1	57.8
1968	-111.3	3.8	-18.0	40.8	83.3
1969	-209.6	45.0	-33.2	100.7	95.7
1970	-254.7	62.4	-41.2	127.0	104.2
1971	-168.5	4.3	-23.1	66.8	119.1
1972	-434	95.8	-67.5	223.0	160.8

Source of data: Mendes (1985a).

loans only the interest rate subsidies received under the system should be treated as effects of integration, and these were rather small. Between 1970-78 they were, on average, around 10 million ECU's per annum. But, from 1979 on they were substantially increased due to the EMS, attaining a value of 251.1 million ECU's in 1980 (European Economy no.13 1982).

Another difficulty arises from the fact that even in the country break down of the contributions to the financing of the budget, only own resources are taken into account, thus excluding some contributions which had a significant value in 1980 and 81.

An item which could be split between participating countries is the contribution to the EDF; however, because this represents small amounts we will assume that these contributions to less-developed countries would have been made anyway on a bilateral basis.

Finally, it should be noted that the values obtained are likely to be different from those registered in balance of payments accounts of member countries, but as long as they show the same pattern of growth our results will not be seriously affected. An illustration of these differences is provided by the U.K. (where such data are available) for the period 1973-81 when the following budgetary transactions regarding balance of payments were registered:

1973	1974	1975	1976	1977	1978	1979	1980	1981
-62.3	-29.6	6.7	-139.9	-300.7	-618.8	-700.1	-498.5	-294.8

Source: United Kingdom - Balance of payments, 1982 ed. HMSO Table 12.2 pp.64 - values in million ECU's at current prices.

while the EEC accounts show quite different values (see table 5.10 below).

The net effects on the balance of payments are presented in tables 5.9 and 5.10 for the periods 1958-72 and 1973-81.

From table 5.10 below it is evident that, apart from the CAP, the common budget has no significant redistributional effects. In fact all countries contribute with the exception of Belgium-Luxembourg (where the institutional set up of the Rome Treaty was located), Ireland (a small amount) and the U.K. (only in last two years).

Table 5.10 - Annual Net Flows From the Budget 1973-81
in million ECU's at current prices

		1973	1974	1975	1976	1977	1978	1979	1980	1981
Germany	Total	-577	-628	-556	-988	-704	-968	-1586	-1593	-2461
	other	-354	-419	-255	-364	-218	-423	-706	-630	-1379
	CAP	-223	-209	-301	-624	-486	-545	-880	-963	-1082
France	Total	-219	-316	-60	-81	-98	-583	-190	435	59
	other	-356	-411	-284	-306	-156	-241	-364	-218	-836
	CAP	137	95	224	225	58	-342	174	653	895
Italy	Total	-237	-194	26	-134	-136	331	449	768	393
	other	-237	-268	-147	-92	123	-89	126	330	-165
	CAP	0	74	173	-42	-259	420	323	438	558
Netherl.	Total	-170	-179	-132	121	139	4	232	428	39
	other	-114	-135	-78	-99	-49	-118	-202	-155	-328
	CAP	-56	-44	-54	220	188	122	434	583	367
Belg-Lux	Total	39	122	180	295	418	539	630	529	520
	other	142	221	303	316	427	573	580	669	650
	CAP	-103	-99	-123	-21	-9	34	50	-140	-130
U.K.	Total	-20	-82	-189	-239	-67	-336	-937	-1324	-703
	other	-108	-107	-55	-68	88	-51	-30	126	586
	CAP	88	25	-134	-171	-155	-285	-907	-1450	-1289
Ireland	Total	43	51	78	192	423	350	530	693	565
	other	-7	2	1	24	47	50	121	190	168
	CAP	50	49	77	168	376	300	409	503	397
Denmark	Total	82	85	106	240	282	330	370	338	226
	other	-25	-25	-32	-5	-6	-33	-25	-40	-73
	CAP	107	110	138	245	288	363	395	378	299

Source of data: Mendes (1985a).

Note: Net positions do not sum to zero due to expenditures made in third countries and carry-overs from preceding years.

5.4 - Foreign investment and labour remittances

To complete the estimation of integration effects on the balance of payments let us turn now to an assessment of factor movements (capital and labour). It is convenient to start by recalling that, as stated in Chapter II, traditional customs union theory assumes that factors are immobile. Nevertheless, ever since Ohlin (1933) pointed out that international factor mobility is a substitute for commodity trade, it has been acknowledged that tariffs and other commercial policies do have effects on international factor movements (Balassa 1976). However, the theory of international capital movements and migration and their relation to economic integration is not well developed. Since this aspect is outside the scope of this work, we shall utilise the customary approach of an eclectic evaluation of the sparse evidence available. To start with some boundaries must be defined a priori: (1) both capital and labour movements raise important political and economic issues which might be even more important than the direct effects on the balance of payments considered here; (2) all the policies of the EEC institutional set up extending a simple customs union will not be considered explicitly. In fact, some were only implemented late in the period considered, namely the European Monetary System (EMS), and some are simply planned such as the 7th Directive on consolidated company accounts. In particular, this is one reason why the analysis here will be limited to direct foreign investment, leaving aside portfolio investment and short-term capital movements.[9]

Capital Movements (Direct Foreign Investment)

There are several theories of direct foreign investment of which Dunning's (1977) eclectic hypothesis seems to be better suited to our interests because it gives joint consideration to the causes and choice of location issues. In particular, none of the other theories seem to explain, in a clear-cut way, why, within the EEC, the German investors prefer to invest in France, Belgium and Luxembourg, while the Americans prefer to invest in the U.K. and Germany, and whether integration does affect the specific locational factors of each country.

In general, a simple look at the time series of net direct investment recorded in balance of payments accounts suggests that EEC integration might have some influence on its evolution (in spite of the usual large fluctuations of capital movements). A closer look at individual countries such as was done for the U.K. (Mayes - 1983) and Germany (Paine - 1979), also points in the same direction and suggests that investors' reactions might involve both leads and lags of around two years.

As far as the EEC is concerned, the literature has concentrated on the testing of the so-called tariff discrimination hypothesis. This was rigorously formulated by Mundell (1957) in a 2x2x2 model and states that capital movements are a perfect substitute for product trade. Therefore, foreign investment is undertaken in countries where it is relatively difficult to export due to the existence of tariffs so that as a result of trade liberalisation one would expect an increase in exports (relative to trend) and a reduction in international investment between partners and vice-versa with non-partners. In the case of the EEC, for example, one would expect USA investment to increase due to the common external tariff (and correspondingly exports into the EEC reduced) while German investment in the USA would increase faster than that in its EEC partners due to the freeing of trade within Europe and the existence of trade barriers in the USA (and conversely in the case of exports).

In fact, we cannot expect this substitutability hypothesis to hold in all circumstances, and, as Schmitz-Helmbergen (1970) have shown trade and factor movements may be complementary rather than substitutes. If the basic model is amended to incorporate inter-country differences in natural resource endowments, so that we can consider a primary manufacturing sector and a secondary manufacturing sector, then capital movements and product trade (especially in primary goods) may complement each other. This is important because it is then possible to accommodate empirical findings pointing to external trade creation relatively to the U.S. (see section 5.1 above) within the tariff discrimination hypothesis. Nevertheless, it should be noted that only in two cases (the Netherlands in 1977-79 and the U.K. in 1979-81) was this verified. The major exception to the tariff discrimination hypothesis was the case of Ireland which suggests that investment was basically driven by abundant and cheap labour. A similar pattern was also found in relation to U.K. investment in Ireland.

It is also important to summarize some of the effects of direct foreign investment which include increased exports to the host country covering capital and intermediate goods, as well as services required by market expansion. An increase in exports by the host country is also possible, especially if the investment is orientated towards the exploitation of cheaper production factors (which is not the usual case in integration induced investment), as well as a reduction in imports which are substituted by sales of the new enterprises created. All these investment related trade flows cannot be estimated since, because of the residual technique used, they are mixed up with the estimation of the integration trade effects presented in section 5.1.

Other foreign investment effects with an impact on the balance of payments, such as the payment of patents, royalties and the transfer of profits, will also be excluded due to accounting difficulties. Attention will focus only on annual flows of new direct foreign investment. For reasons suggested above, these operate both ways and the net outcome cannot be predicted a priori; rather it is an empirical question for individual countries.

Turning again to the tariff discrimination hypotheses, and why integration affects investment flows, we know that apart from the trade barriers the rise of the new (enlarged) market also seems to be one of the key determinants of foreign investment. With a smaller role we could also expect the growth (induced by integration) of the market to play a part in the process of attracting foreign investment as well as in the harmonization of rules and patterns and in generating an increased supply of factors.

So far the empirical testing of the hypothesis [eg. Schmitz-Bieri (1972); Lunn (1979;1983); Scaperlanda-Balough (1983)] has concentrated on U.S. direct investment[10] and until the early 1970's the overwhelming result was a rejection of the hypothesis. Since then, however, the hypothesis has been accepted. The first to claim the validity of the hypothesis were Schmitz-Bieri (1972) who, using

Table 5.11 - Estimated net flows of direct foreign investment induced by integration

in million ECU's at current prices

Years	Germany	France	Italy	Netherl.	Belg-Lux.	U.K.	Ireland	Denmark
1974	192.8	-33.2	316.7	0	-76	399	-55.9	-356.5
1975	450.3	84.4	282.0	78.5	-99.5	751.4	-153.8	-392.4
1976	863.9	61.6	146.2	49.6	-159.8	267.7	-225.8	200.3
1977	1075.2	-68.4	206.5	-86.6	-181.5	346.1	-277.9	81.6
1978	1450.4	-158.8	404.6	-198.8	-46.6	678	-308.9	-369.3
1979	2045.4	15.7	541.1	-877.8	-114.5	712.5	-344.5	-521.2
1980	2293.7	456.7	566.1	-1398.7	-215.8	1484.8	-266.2	-199.6
1981	2436.9	347.4	490.7	-2210.2	-629.3	769.3	-109.0	156.1

Notes: - For inflow and outflow values see Mendes (1985a).
- A negative sign means a positive effect on the balance of payments

both a simple trend model with dummy variables and a demand model, concluded that up to 1966 integration in the EEC had significantly attracted U.S. investment. Their dependent variable was the share of U.S. investment within the EEC which was explained by the EEC Gross National Product, lagged one year, and its growth rate, plus the EEC share of total U.S. exports (which is the test variable for the tariff discrimination hypothesis) and a multiplicative dummy. Contrary to previous research, all the variables were significant at the 90% confidence level and had the expected sign. Their results were criticised mainly on the grounds of inappropriate variables and data. The central problem is that in view of the possibility of complementarity and some evidence of external trade creation, the share of U.S. exports to the EEC is not appropriate to account for the discrimination hypothesis. However, further research by Lunn (1979;1983) and Scaperlanda-Balough (1983), extended the basic model by introducing new variables to account for U.S. foreign investment control programmes and lagged net fixed assets of foreign affiliates, plus the use of better specified variables: their results confirmed the tariff discrimination hypothesis. In particular, the results by Scaperlanda-Balough are very significant, not only because Scaperlanda was one of the most respected scholars to refute the hypothesis, but also because they used a proxy for the discrimination hypothesis based on a dummy variable scheme, representing the progressive dismantlement of industrial tariffs on intra-EEC trade, and a longer time series. Therefore, the problems of substitutability-complementarity were by-passed and their results are more reliable. They concluded in favour of the prevalence of the hypothesis, confirmed the greater importance of the market size variable and suggested that after 1972 account should be taken of

Table 5.12 - Integration Effect on Foreign Investment of the United States

in million ECU's at current prices

		Germany	France	Italy	Netherl.	Belg-Lux.	U.Kingdom	Ireland	Denmark
1974	from	-9.6	22.7	-6	-1.3	-.5	62.6	0(1)	0(1)
	to	-27.3	-39	32	-53.1	-69.4	-45.4	8	3.3
1975	from	73.5	-16.8	-.3	24.8	-90.9	208	0	0
	to	-76.8	-152.8	35.8	-38.3	-140.4	26.0	31	11.3
1976	from	24.5	-39.6	-31.8	19.9	-113.4	378.6	0	0
	to	14.1	-271.7	-50.2	-30.3	-120.3	109.6	63.3	7
1977	from	148.6	-91.1	-28	93.9	-155.1	615.7	0	0
	to	170.2	-356.8	64.4	90.6	-182	236.1	106	7.8
1978	from	226.2	-107.3	-41	77.2	-208.6	1332.8	0	0
	to	62.3	-359.1	144.1	142.2	-76.2	419.4	56	3.7
1979	from	68.3	-55.6	-44.7	128.1	-173	2203	0	0
	to	334.5	-270.5	285.2	7	-212.6	1149.5	94	13.3
1980	from	196.8	64.3	-62.7	279.5	-110.9	3210.6	0	0
	to	148	-429.1	130.5	-93.7	-278.3	802.5	153	21.7
1981	from	144.9	357.9	18.2	510.9	44	2825.8	0	0
	to	173.6	-493.4	-15.9	-171.1	-530.9	490.2	301.3	16.3

(1) Values are always zero because there was no split of data of investment in the USA by Ireland and Denmark.

the flexible exchange rates system. This evidence, together with the observation of the time series referred to above, precludes us from accepting the customs union assumption of factor immobility (or its minor relevance) and requires that their balance of payments effects be measured.

However, the acceptance of the tariff discrimination hypothesis, although indicating that foreigners increase their investments in the EEC and that this increases relatively its investment abroad, does not explain why firms locate in particular countries or why their individual net outcomes are not estimated. As an exploration of these aspects is required in our approach, one possibility of by-passing this difficulty would be the use of a trend model of the kind referred to above with net flows taken from the Balance of Payments accounts or from a more disaggregated data base. An alternative that seems more plausible is the use of the weighted share analysis presented above; this is possible if sufficiently[11] disaggregated data exists for at least a few years. In fact, there is a parallel with the trade flows if one takes as the price of capital its return in each country. We would have an elasticity of substitution between competing sources of supply (demand for) of capital, and a share index can be defined in a similar way. Here the relationship between tariff changes and rates of

Table 5.13 - Integration Effect on Foreign Investment of the United Kingdom

in million ECU's at current prices

		Germany	France	Italy	Nether.	Belg-Lux.	Irl.(1)	Dk.(1)	EFTA
1974	from	2.3	-2.9	5.4	9.2	9.2	-1.3	-295.2	33.3
	to	-21.1	-32.1	-5	-48	-22	31.6	9.3	-26.5
1975	from	7.3	18.2	1.9	-2.7	2.7	5.8	-470.1	71.2
	to	-17.1	-74	3.7	-50.1	-26	72.6	-53.9	35
1976	from	1.1	44.3	-2.1	48.4	23	10.5	112.9	105.7
	to	20.8	-73.6	2.7	-55.4	-26.6	103.7	-45.1	86.8
1977	from	15.8	96	-15	31.4	25.4	14.2	-55.8	97.7
	to	13.3	-134.9	25.1	12	-51.6	157.4	-71.4	187.2
1978	from	22.8	47.8	-26.7	73.2	10.3	4.9	-367.8	162.6
	to	-119.1	-189.1	30.4	-88.4	-111.3	183.8	-30	94
1979	from	10.5	-29	-39.9	168.3	-.3	2.2	-486.8	159.6
	to	-11.7	-153	40.8	-151.1	-167.7	198.1	-32.4	-6.8
1980	from	72.6	-127.9	-49.5	160.6	-4.5	-1.3	-197.7	325.3
	to	-27	-177.3	32.9	-203.5	-113.4	129.6	-42.9	-162.1
1981	from	71.2	-150.3	-58.4	220.6	39.1	6.8	157.5	322.2
	to	27.2	-170.4	8	-184.8	-143.5	40.8	-31.9	-105.5

(1) - The annual values are substantially downward biased due to large disinvestment in the "rest of the world" estimated from insufficient data.

return in different countries is not so straightforward but we can assume a simple relationship of the form presented in chapter IV (expression 4.27).

Estimates for the period 1974-81, presented above in table 5.11, were obtained using the share technique. Poor data and the simplest form of share analysis (aggregation, no home market and interdependence bias) impose serious limitations on the results although they do not look awkward in terms of sign. It can be observed that only the smaller countries (Ireland, Belgium-Luxembourg, Netherlands and Denmark) improved their balance of payments. This result, in the case of the new members Ireland and Denmark, conforms to what one would expect given their small amount of investment abroad and substantial investment attracted from the United States (see table 5.12 above), although in the case of Denmark it is mostly due to a higher disinvestment in the U.K. (see table 5.13 above).

It is in fact interesting to match this last effect against the opposite one verified between the U.K. and the remaining members of EFTA. Here, following the free trade area with the EEC, it seems that due to fear of increased competition from other EEC partners, both the U.K. and EFTA increased investment on a bilateral basis. These results clearly conform to the tariff discrimination hypothesis.

For the Netherlands[12] we estimated a significant positive effect in spite of an increased investment in the U.S. and U.K. which nevertheless was not enough to offset the changes in investment with the "rest of the world" (other EEC included) and investment from the U.S.. On the other hand, Belgium and Luxembourg had their inflows and outflows reduced, which, coupled with a large inflow from the "rest of the world", produced a net positive effect.

Finally, both tables give strong support to the tariff discrimination hypothesis with the U.S. consistently investing in U.K., Ireland and Denmark and the U.K. investing in the U.S. and EFTA. To reinforce this conclusion we can add the fact that U.K. also reduced its investment in the other[13] member countries of the EEC.

Labour Movements

Turning now to the labour movements one would expect migration between member countries to occur mainly as a result of the elimination of administrative barriers and some harmonization of the welfare state of individual countries together with a community recognition of professional qualifications and degrees. However, there is little explanation of the impact of the EEC on these issues and concomitantly there is still less information on the savings and transfer behaviour (remittances) of the migrants. Furthermore, the theory of international migration does not provide enough explanation on the role played by institutional and linguistic factors (or "psychological distance") that work as deterrents of migration, offsetting the inducements to migration that one could expect, in a more integrated labour market, from earnings and employment differentials.

The general picture one gets from tables like 5.14 below is that labour movements are very limited, except in a few countries with particular characteristics (Italy and Ireland) and are not strictly related to membership of the Community.

If one examines data on earnings and unemployment differentials, together with the well-known inflow of migrants from the periphery in the 1960's and early 1970's, it is striking that, in spite of the existence of the Community, the level of migration is very low. Perhaps we ought to accept Mayes' (1983) conclusion that the "major reason for this is the lack of integration".

Nevertheless, while it is acceptable that Ireland's particular relations with the U.K. in terms of labour market have not been affected by the EEC, it must nevertheless be regarded as plausible that Italy might have benefited from membership, both in terms of in-migration relatively to the European periphery during the 1960's and in terms of a less severe reflux in the second half of the 1970's.

We will test the hypothesis that Italian migration benefited from integration by using a simple trend model for the remittances into

Table 5.14 - Migration between EEC Countries

Countries	Nationals working in other member countries, 1976		Earnings and Unemployment indices Average of the years 1970 and 1975	
	thousands	as % of domestic labour force	Earnings * (base - highest=100)	Unemployment rate (base - lowest=100)
Germany	137	0.5	75	100
France	114	.5	50	138
Italy	694	3.6	46	236
Netherlands	83	1.8	68	104
Belgium-Lux.	68	1.8	68	128
United Kingdom.	61	.2	55	119
Ireland	455	44.6	46	260
Denmark	7	.2	100	119

Sources: Mayes (1983) table 6.10; Paine (1979) table I and OECD - Labour Force Statistics 1968-79

Notes: * - Average base hourly wage costs are used as a proxy for earnings.

Italy and other similar Mediterranean countries for the period 1970-81.

Although the results presented in table 5.15 below are statistically poor there is no doubt that they support the above hypothesis. In fact, all countries experienced a substantial reduction of remittances during the period 1973-76, due to the world crisis, but only Italy presented a (relative) recovery after 1977 while at the same time all the other countries were facing restrictions in entering the EEC labour market and their workers were returning home in large numbers.

The fact that community employers and governments were not able to discriminate against Italian workers became a benefit for Italy which stemmed from integration. Therefore, in calculating the balance of payments benefit for Italy we will take the amount corresponding to the difference between the estimated values and those that would have been verified if the 1970-76 trend had continued, less the percentage verified in non-member countries. The total annual amount was:

Table 5.15 - Estimated regression coefficients for remittances
period 1970-81

countries	constant a_0	trend a_1	intercept a_2	slope shift a_3	R^2	D.W.
Italy	7.652*	-.213*	-1.067**	.197*	.85	1.47
Greece	6.975*	-.075**	.722	-.070	.71	1.71
Spain	7.378*	-.074	.473	-.037	.24	2.18
Portugal	7.421*	-.089**	-.661	-.072	.78	1.06

Notes: * - significant at the 99% level; ** - at the 90% level; otherwise non significant.

- The model is of the form $\ln X_t = a_0 + a_1 T + a_2 D + a_3 TD + u$, where X is the remittances; T is time, and D is a dummy variable with D=0 until 1976

1977 = 208 ; 1978 = 275 ; 1979 = 331 ; 1980 = 377 ; 1981 = 417 (million ECU's)

We are now left with the question of whether such an inflow should be accounted as an outflow for the host countries and consequently taken as a cost of integration. Although this seems to be a logical step in terms of balance of payments, there are, however, a few arguments which suggest that it is not this simple. Firstly, the crude nature of the estimates and the rationale behind using any "ad hoc" procedure to split up those values by source (mainly Germany and France) suggest we should not consider them as a cost. Secondly, the amounts would be rather small in any case. Thirdly, migrant workers would have been substituted by less efficient indigenous unemployed which would have been paid the same (or even higher) wages and would be likely to have higher propensities both to import and to consume. The consequent losses in efficiency and rise in imports would presumably be high enough to offset the remittance outflows and, therefore, it seems reasonable to assume that the costs of "imported" labour do not represent a balance of payments cost for host countries.

Having completed the estimation of the most important direct effects of integration on the balance of payments, which result from trade liberalization in industrial goods; a common agricultural policy; a common budget and factor mobility; we will now proceed to introduce them into the balance of payments approach developed in chapter III so that integration and growth performance can be evaluated.

1. It can be easily demonstrated from equation 4.43) that this result depends on the following inequality: $(a^{'}-a)\sum_{s=1,s\neq i}^{m}\Delta_{sj}+(b^{'}-b)\sum_{s=1,s\neq j}^{m}\Delta_{is}>a\Delta_{k_j}+b\Delta k_i$ with $a^{'}$ and $b^{'}$ equal to $\frac{a}{1-a-b}$ and $\frac{b}{1-a-b}$ of 4.43) without the home market. As we have shown that $(a^{'}-a)$ and $(b^{'}-b)$ are both greater than zero and the condition will easily be met as long as the summations of all other countries' changes in external trade are in general much greater than an individual country's change in its home market.
2. These were: 6 - manufactured goods; 7 - transport equipment, and 8 - miscellaneous manufactured goods.
3. Most of the political quarrel about the budget within the community is still largely limited to these, as has been widely illustrated by the press coverage of the Athens (1983) and Brussels (1984) summits.
4. In fact this designation, used in some studies such as the one of Rollo and Warwick (1979), is less appropriate as it excludes the first effect mentioned - changes in international trade flows.
5. Blancus (1978) used world prices as reference but did not attempt to measure volume effects.
6. See Rollo-Warwick (1979) on this question.
7. These results are comparable to those of Warwick and Rollo (1979) except in the case of Germany where they found a negative effect.
8. Therefore the following estimates should not be regarded as exact but only as giving an order of magnitude.
9. For a taxonomy of international capital theories, see Grubel (1982)
10. Appropriately this should cover only the manufacturing sector, but a few studies consider total investment.
11. Obviously the usual caveats about data on financial flows (Laffargue - 1974) and drawbacks of the share approach are much more stringent here.
12. Its improvement from 1977 onwards is anyway possibly due to North Sea oil related investments.
13. The exception shown by Italy is possibly due to investment in sectors other than manufacturing, such as real estate and tourism.

Chapter Six

THE CONTRIBUTION OF EEC TO ECONOMIC GROWTH

6.1 - Introduction

We have now estimated the various integration induced effects required to use the balance of payments approach developed in chapter III, in order to answer some of the questions towards which our research project is directed. However, some preliminary remarks and qualifications are needed in order to explain the results presented in the forthcoming sections. First, it must be remembered that the framework developed relies on disregarding the price volume effects referred to in section 3.2 which, in any case, were shown to be rather small, but we could not bypass the problem of interdependence in the estimation of changes in the terms of trade and growth rates. Second, owing to the extensive data requirements involved, the contribution of integration to the growth rate could not be estimated as the sum of the partial effects but, instead, it had to be based on the effect of integration on both imports and the income elasticity of import demand. It therefore follows that the framework is more appropriate for distinguishing the various mechanisms through which integration affects the growth rate of member countries, rather than providing estimates of its exact magnitude. A third, and more serious drawback, arises because the approach relies basically on trend growth rates, with the consequence that the results are largely dependent on the possibility of obtaining significant trend growth rates and the choice of the time horizon during which we expect the integration effects to prevail. However, these are not insoluble problems and we have used a simulation algorithm in order to generate the probability distribution of possible outcomes.

Other minor problems arise from the fact that estimates of the CAP effects are (upward) biased due to the use of total trade in food and beverages, which includes both manufactured and raw food stuffs, while only the last items are covered by the Common Agricultural Policy. Furthermore, as the effects on trade were estimated before allowance was made for the effects of increased prices in food commodities there is also a small amount of double counting.

One should also remember the problems (vide previous chapter) associated with the estimation of the home markets. Nevertheless, these are mostly of an operational nature and could be bypassed given time and the resources needed to use the "ideal" data required by the approach. There is, however, an important and basic problem which remains unsolved. It relates to the method by which the likely relative shrinkage in the home market, resulting from increased specialization and external trade, should be incorporated into the framework. In fact, experimentation (not reported here) with the national accounts values for the nineteen-sixties suggested that this shrinkage might be substantial and therefore the results presented below must be seen as overestimates. But, all in all, there would still be very significant effects of integration upon the growth rate and the general picture depicted from the results presented would not be reversed.

These are the general qualifications; we will state the more specific ones at the beginning of the respective sections. We now proceed to the presentation of the results relating to the EEC performance, following which we will draw some policy conclusions.

6.2 - Integration in the 1960's - The EEC, period 1961-72

Estimates for this period of European integration were based on total trade data which did not allow for a complete breakdown of integration effects. Moreover, it had the consequence that aggregation bias, both at the commodity and at the country level, was much larger (see previous chapter). In particular, and as a result of this aggregation, the estimates for France are unreliable because most of the large trade diversion it experienced (see previous chapter) was with the "rest of the world" country which certainly includes the loss of special links with former colonies rather than integration. Further disaggregation to eliminate this effect would substantially reduce the negative effect estimated. It also happens that aggregation does not take into account the fact that during most of the 1960's integration only affected manufactured products and therefore trade changes in both food and raw materials introduce a substantial bias in the results.

Furthermore, the values presented for foreign investment are not based on any direct estimate but on some assumptions regarding the balance of payments effects of integration. In fact, it was assumed that there were no capital flows other than those generated by the EEC budget, and that the terms of trade would remain constant, so that foreign investment was estimated as the difference between the integration effects on imports (times the terms of trade minus one) and net budget flows. Therefore, it includes the interdependence factor plus any other capital flows such as those resulting from labour remittances (see expression 3.49).

Turning now to table 6.1 below we see that the most important finding is the large effect that integration has had on growth. In fact, even if we admit that the above mentioned overestimation was 100%

Table 6.1 - Integration effects on the % trend growth rate of member countries
- EEC5 1961-72 -

Contributions to growth	Germany	France	Italy	Netherlands	Belgium-Lux.
Actual growth rate	4.39	5.40	4.97	5.17	4.56
1 - Growth rate due to EEC [1=-2-3+4+5+6+7+8]	-0.02	-2.71	1.04	2.94	2.45
Which was made up of					
2-Terms of trade changes	0.02	0.57	0.94	0.19	-0.17
3-Change in propensity to import	2.25	2.90	1.12	0.44	1.16
4-Growth of export volume	3.52	1.25	4.74	4.09	4.56
5-Change in the trade balance position	-0.45	-2.08	0.05	-0.06	-0.71
6-Net EEC budget payments	-0.14	0.01	-0.28	0.12	0.09
7-Foreign investment	-1.02	-0.05	-1.13	-0.15	-0.07
8-Residual + errors [8=(1+2+3)-(4+5+6+7)]	0.34	1.64	-0.28	-0.43	-0.43

Notes: As defined in the model a negative sign of the terms of trade means an improvement;

-The total effect may differ from the sum of its components due to rounding.

we are still left with values slightly higher than 1% point. These are values much higher than suggested by previous studies [eg. Balassa (1975)], and for the most favoured countries (Netherlands and Belgium-Luxembourg) integration accounts for around 55% of the actual growth rate experienced by these economies.

Furthermore, although during this period other important tariff reductions were agreed within the GATT arrangements [the Kennedy (1964) and Dillon (1968) rounds of tariff reduction] there is no doubt that the bulk of the growth achievement must be attributed to the formation of the European Economic Community.

The second most striking result is that those countries that did not benefit from integration were precisely the two major economies. Germany suffered a very slight loss and France apparently suffered a major loss (although this might be overestimated for the reasons advanced above). This result contrasts (particularly in the case of Germany) with a large trade creation effect, (see Table 5.2), in spite of some trade diversion relatively to the United States.

However, as one would expect from the opening of individual markets there was also an offsetting increase in the propensity to import which reduced the growth resulting from increased export growth. These two countries were the ones with the largest increase in the propensity to import. They were the only ones to have a significant positive shift of their income elasticity of demand for imports which was large enough to meet the condition mentioned in chapter III (ie $\Delta\pi > \frac{\pi}{m}\Delta m$, with $\Delta\pi > 0$) in which a country will experience a loss in spite of having benefited from trade creation. These results clearly confirm our claim that it may be misleading to rely on estimates of trade creation alone to assess who gains from integration when the question is looked at in a growth framework.

It must also be emphasized that during this period the most important contributions to the growth rate came from the growth of exports. Other mechanisms such as the budget or foreign investment played a relatively small part in the process. The only country to be significantly affected by budgetary transactions (but negatively) was Italy, in spite of being the poorest member.

As far as the terms of trade are concerned Belgium-Luxembourg was the only country to experience a significant positive contribution resulting from improvement in the terms of trade. In all of the other countries there was a negative effect. Note that this effect is measured against a background (referred in chapter III) of a generalized improvement in the terms of trade over the period. This result contrasts with the suggestions given by Petith (1977) that improved terms of trade could be the single most important gain of integration in the EEC context. However, his other suggestion, that under his assumptions[1] the smaller the country the larger the terms of trade gains, seems to be confirmed during this period. Finally, it may be noted that the poorest/weakest economy (Italy) was the one which experienced the largest loss.

As to the effects resulting from increased mobility of factors, grossly called foreign investment[2], we see that all countries were negatively affected, although this was only significant in the cases of Germany and Italy. But, given that there was only a large outflow of capital in the case of Germany it is logical to admit that the Italian position might be largely due to the interdependence effects between the terms of trade and imports.

6.3 - The EEC after enlargement - period 1974-81

For this period, a larger and more disaggregated set of data were used which increases the reliability of the estimates. The only exception is that of the Netherlands where the problems relating to the estimation of the home market may have introduced a substantial bias. There is, however, a qualification that must be made before proceeding to the presentation of the results. It concerns the option of using all trade effects, and not just those referring to trade between new and older members. Apart from minor arguments, such as those concerning the possibility of trade reorientation, the central reason is that during the period of analysis there was a general upsurge in protectionism. This had two main consequences; one concerning the increase in the EEC's own level of protection, mainly through non-tariff mechanisms (that could be transformed into their tariff equivalent), which can be interpreted as a rise in the common external tariff, and the other (probably the most important) concerning the increase in protectionism elsewhere in the world, particularly by other OECD countries.

This last phenomenon, which can be compared to a case where the "rest of the world" is retaliating against the customs union, either by increasing its own level of protection or by forming its own customs union, has not been dealt with in the theory and its consequences in terms of global trade creation and diversion are not clearly identifiable. However, it is plausible to accept that faced with increased competition outside the union, the EEC countries were likely to react by increasing the share of intra-union trade. In fact, the estimates presented in the previous chapter confirm that the increased intra-trade between the former members was quite substantial which gives support to this hypothesis. It therefore follows that integration effects in the period 1974-81 are a consequence both of the enlargement of the EEC and of increased protection on a worldwide scale.

Turning now to the results presented in table 6.2 below we can observe that the effects of integration continue to be quite substantial and mostly positive.

Denmark, one of the new members, is the only country that has apparently lost from integration mainly because it did not have a significant amount of trade creation, which even showed a declining trend (see Table 5.3). France in contrast to the earlier period was the country which fared best in both absolute and relative terms, with over half its actual growth rate accounted for by integration effects. Comparing the performance of the new members with that of the former members, the new benefited least in absolute terms relatively to all the older members, although in relation to the actual growth rate the U. K. apparently experienced a larger benefit than the Netherlands and Italy. This performance deserves special mention since large sectors of public opinion in the U.K. believe that integration has worsened the rate of growth through adverse balance of payments effects. But in spite of adverse effects of CAP the results show that integration accounts for about 30% of growth during this period.

If one looks at the breakdown of integration effects, the most important driving force was still the growth of exports, although in this period it was not so dominant. Furthermore, an important part

Table 6.2 - Integration effects on the growth rate of member countries

- EEC8 1974-81 -

Contributions to growth	Germany	France	Italy	Netherlands
Actual growth rate	2.65	2.66	2.74	1.99
1 - Growth due to EEC [1 = -2-3+4+5+11+...+16]	0.91	1.57	0.42	0.53
Made up of				
2-Terms of trade changes	-0.01	-0.39	0.05	0.07
3-Change in propensity to import	0.88	0.79	0.40	0.29
4-Growth of exports of manufactures	0.02	0.30	0.07	2.33
5-Change in the trade balance of manufactures	0.74	0.66	-0.83	-2.66
6-Growth of exports of food products	0.50	1.35	0.14	1.28
7-Change in the trade balance of food	-0.44	-1.20	-0.35	-0.47
8-Export gains due to increased prices of food	0.02	-0.07	-0.10	-0.63
9-Change in the balance of price effects	-0.24	0.08	0.09	0.07
10-Net CAP - budget payments	-0.05	0.08	0.06	0.26
11 - total CAP effects [11=6+7+8+9+10]	-0.21	0.23	-0.15	0.51
12-Net non-CAP budget payments	0.06	0.02	0.20	0.03
13-Labour remittances	..	..	.08	..
14-Direct foreign investment	0.35	0.03	-0.04	-2.53
15-Interdependence effect	-0.25	0.01	0.47	1.71
16-Residual + errors [16 = (1+2+3)-(4+5+11+...+15)]	1.06	0.72	1.06	1.49

Notes:

- As defined in the model a negative sign of the terms of trade means an improvement;

-The total may differ from the sum of its components due to rounding;

- Appropriately the CAP total should also include effects on the terms of trade and propensity to import. The total CAP effects are therefore likely to be upward/downward biased.

- Trade estimates for the Netherlands are substantially upward biased for reasons referred to in the previous chapter (entrepot trade and home market).

Table 6.2 (cont.) - Integration effects on the growth rate of member countries
- EEC8 1974-81 -

Contributions to growth	Belgium-Lux.	U.K.	Ireland	Denmark
Actual growth rate	2.03	1.24	3.84	1.98
1 - Growth rate due to EEC [1 = -2-3+4+5+11+...+16]	0.71	0.37	0.31	-0.64
Made up of				
2-Terms of trade changes	-0.22	-0.31	0.32	−3.09[1)]
3-Change in propensity to import	0.57	0.32	3.28	0.21
4-Growth of exports of manufactures	0.27	0.79	2.38	-0.08
5-Change in trade balance of manufactures	0.74	0.50	-1.72	-2.86
6-Growth of exports of food products	0.42	0.65	1.12	0.78
7-Change in the trade balance of food	0.01	-1.98	-0.50	-0.46
8-Export gains due to increased prices of food	-0.13	-0.03	-0.38	-0.17
9-Change in the balance of price effects	-0.12	-0.04	0.36	0.12
10-Net CAP - budget payments	0.06	-0.32	0.86	0.19
11 - total CAP effects [11=6+7+8+9+10]	0.24	-1.72	1.47	0.45
12-Net non-CAP budget payments	0.09	0.08	0.54	-0.04
13-Labour remittances	..	..	..	..
14-Direct foreign investment	-0.02	0.09	0.20	0.89
15-Interdependence effect	0.19	-0.48	1.37	-0.73
16-Residual + errors [16=(1+2+3)-(4+5+11+...+15)]	-0.45	1.14	-0.32	−1.58[1)]

Note: 1) - Estimated without considering total capital flows in expression 3.44

of this effect was the growth of exports in the food and beverages sector. In this period, as expected, the offsetting effect of an increase in the propensity to import was much smaller, with the exception of Ireland. In this regard Ireland contrasts with the former EFTA members (U.K. and Denmark) which had a much smaller change in the propensity to import. The explanation lies in their existing membership of EFTA and the associated links between these two European groupings.

The terms of trade effects, as in the previous period, were also small, although the sign is now generally reversed. Only Italy, the Netherlands and Ireland suffered a loss while all the others experienced a positive gain. Also, in contrast with the nineteen-sixties, this occurred during a period when the global terms of trade were deteriorating. On this account, there is some (weak) evidence to suggest the hypothesis that an economic union is more likely to secure gains through terms of trade improvement during a cyclical downturn rather than during an expansion. On the other hand, the view that the weaker members (Italy and Ireland) are the ones more likely to suffer from adverse terms of trade effects is further reinforced. This time the result for Ireland seems to reject Petith's conjecture about the gains to smaller economies, except that Ireland had a much higher initial level of protection which could make the result consistent with his prediction. Finally, the large benefit to Denmark probably reflects the measurement problems referred to in the previous chapter[3].

Turning to factor movements (one of the key stones of the Rome treaty) the results do not suggest that they have had any significant impact on the growth rate. First, the labour remittances, which were only estimated for Italy, had a very negligible effect. Likewise, the effects on growth of the changes in direct foreign investment were also negligible. The only exceptional country is the Netherlands, but here the effect is most likely due to substantial disinvestment abroad and to North-sea oil related investment (see section 5.4 of previous chapter) rather than to integration effects. However, an important difference relative to the nineteen-sixties is that during this period foreign investment generally had a positive effect on growth.

Let us now turn to the Common Agricultural Policy, the only common policy that can be identified separately and treated at a very disaggregated level. The first point under consideration is whether the CAP has been inefficient to the extent of causing a reduction in the growth rate of member economies. The answer cannot be entirely conclusive because it depends on the assumptions made. Assuming that it has not significantly affected the propensity to import and the terms of trade, the conclusion is that, with the exception of the United Kingdom and Germany (and to a lesser extent Italy) all the remaining countries benefited from CAP. Nevertheless, this conclusion depends largely on the hypothesis formulated in section 5.2 that, contrary to previous suggestions, the CAP also had substantial effects on the external trade of foodstuffs. If we had taken the traditional hypothesis and assumed that all the other CAP effects remained the same then we would be left with only the increased price effects, which, although negative, were rather small, and the net contributions to the CAP budget. The net outcome would then depend mainly upon the budgetary transfers and only Ireland, Denmark and Italy would have gained from CAP. It is interesting to note that in this case, where only the budgetary and price effects are considered, Italy shows a positive effect while the U.K. (the big looser from CAP) sees its negative effect substantially reduced to -0.39. The assumption we make regarding the effects of CAP on the pattern of trade is thus crucial for any assessment of its total impact on growth.

Finally, we must consider which countries have their integration outcome significantly affected by the net effect of CAP. Apart from the U.K., which obviously would have a large increase in its integration induced growth rate, some countries (France, Netherlands, Belgium-Luxembourg, Ireland and Denmark) would have their growth rates substantially reduced. However, only Ireland would change its net outcome to a negative overall effect. So, if it were not for the the CAP, integration would have affected Ireland negatively.

We turn now to the effect of other common policies on growth. If one makes the simplifying assumption that they can be reasonably assessed through their budgetary effects then the answer is that they had no significant impact on growth in most countries, with the exception of Ireland and Italy. However, it should also be stressed that with the exception of Denmark (with a very small value anyway) no country seems to be losing on this account. This of course contrasts with the CAP. One is then inclined to admit the possible greater efficiency of other common policies (namely the regional policy) in promoting redistribution without hampering growth and, in some cases, even fostering it.

Finally, an important question is whether the integration effects have been exhausted. As it has been identified that factor movements had a small, and sometimes even negative, effect on growth and that increased budgetary expenditure is dependent upon a stronger political commitment to European unification, we are left with the fostering of exports which, in any case, were the main transmission mechanism of growth. That being the case, and given that the promotion of further trade through CAP mechanisms and protectionism in general are not desirable, then the options left are the arrangements for successive enlargements and the setting up of export promotion policies at a community level. These will be examined in the final section on conclusions and policy issues.

Before summing up the EEC experience it is convenient to point out that, in spite of the large residuals presented, the results are generally reliable. In fact, we performed sensitivity tests for the various variables using a simulation algorithm which allowed us to check the robustness of the results. So, for the period after the first enlargement we can observe (see figures 6.1 and 6.2) that in terms of induced growth of GDP there is substantial variability, but only in one case (Ireland) is the result dubious in terms of a net gain or loss. The Irish values, for an interval of 75%, range between -0.23 and 0.97 percentage points.

Figure 6.1

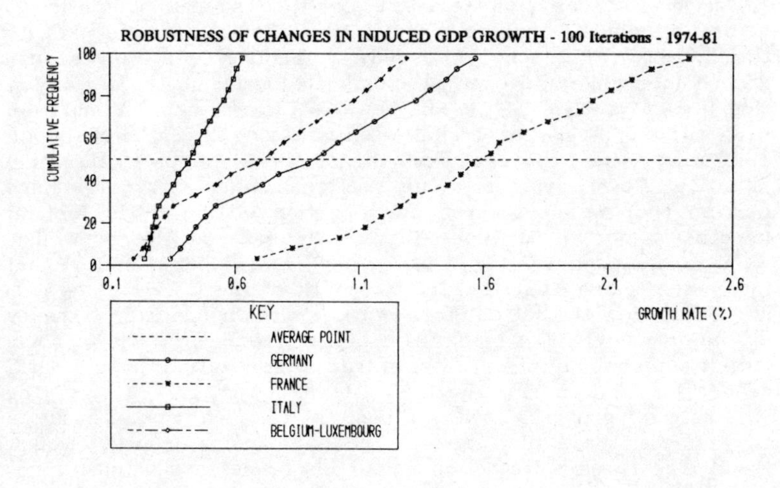

Figure 6.2

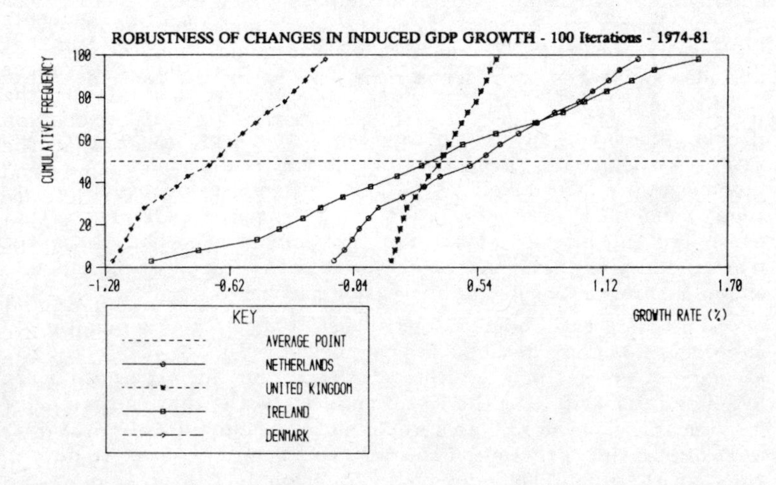

6.4 - Main conclusions and policy issues

In brief, we can summarise by saying that the main conclusion to be drawn from the analysis is that for the past quarter of a century the EEC integration scheme has played a major role in the economic growth achieved by Europe. If the estimates are accepted we can say that in 1972 the GDP of the EEC was 2.2% higher[4] than it would have been without integration, and after enlargement the EEC ended up in 1981 with a GDP that was 5.9% higher than in a non-integration situation. Obviously, this value was compounded over the time horizon that we used given our rejection of the "once for all" hypothesis but, nevertheless, it clearly exceeds the previously widely-accepted guess [eg. Lipsey (1960) and Balassa (1975)] that integration gains would hardly exceed 1% of the GNP. Finally, it is important to note that our framework has confirmed export growth as the driving element behind this process, which in the 1960's was largely supported by the process of trade liberalization through tariff reductions, while during the 1970's it was mostly generated by the enlargement process itself, in conjunction with the association treaties with EFTA and the upsurge of a worldwide protectionist trend. The role of protectionism in this process calls for further research but it can be seen that it rejects the widespread simplistic views based on a clear-cut choice between liberalization and protection and shows that the EEC has a fundamental need to develop an export promotion policy; a view which is reflected in the following discussion. The large gain estimated, however, was not equally shared between all the participating countries. This is not a straightforward claim that the EEC has increased inequality in Europe, because the outcome in this respect varies between the two periods considered and needs careful scrutiny as carried out in section 7.4 below.

In considering other contributors to this growth effect we shall start with the Common Market policies. With the exception of the Common Agricultural Policy, the straightforward conclusion is that their impact in terms of growth was almost nil. However, they presented the interesting feature that almost everybody benefited from them. This underlines the view that the European Community should reinforce its role and increase its scope.

Regarding the Common Agricultural Policy, the examination of its effects was more detailed and complex. First of all, we can not give a clear-cut answer because the gains from the CAP were not shared by all countries. In fact, the losses suffered by the U.K., Germany and Italy meant that the EEC as a whole had a loss in 1981 of 1.7% of her GDP, due to the operation of the CAP. This being the case, we have to question the whole efficiency of the Common Agricultural Policy at a Community level a matter which has, in fact, been at the heart of most of the EEC internal quarrelling, although these quarrels have mainly been limited to budget disputes and not to other issues that our results have shown to be more important. In the light of our results, and using the assumption that everything else will remain constant, we shall then examine the main policy issues that have been present in recent debates.

We shall begin with the drastic policy option of dismantling the CAP. In this case, and assuming that if the member economies were left to evolve their own policies they would not achieve a superior (global) community result, we have that the U.K., Germany and Italy would increase their gain from integration but that conversely Ireland would face a severe loss; Denmark would see her loss increased, and the Netherlands would be in a break-even position. So, unless the U.K., Germany and Italy were prepared to compensate those who would lose if the CAP were dismantled, this step would be unacceptable. However, the supporters of this option can add to their battery of arguments the fact that the CAP has been responsible for the increase in inequality between member countries, in spite of favouring the poorest economy, Ireland.

A second policy option, which has already been adopted to some extent with the budget rebates given to the U.K., would be to maintain the same basic policies but to change the financing of the scheme towards a more equitable system. One approach could be a move towards a greater (or exclusive) financing of the CAP through Value Added Tax contributions, differentiated according to the expected benefits and to the level of income of the participating countries. This is an interesting possibility because it would reduce the losses for the U.K. by almost 20% (less for Germany) but it should be matched against a 40% increase of the Italian losses and a reduction of gains for all the other members, which, in the cases of Ireland and Denmark would be over 50%. Furthermore, this would not be sufficient[5] to eliminate the British loss.

Another policy option (also being implemented in part) involves the implementation of more protection for Mediterranean products which would eliminate the Italian loss. The major problem here lies in the budgetary constraint which, under the present circumstances, would result in an increased burden on the U.K. unless there were a compensatory reduction in subsidies paid on the most favoured northern products.

In fact, an alternative to the scrapping option, which is already gathering some support, would be the reduction of the actual levels of protection for individual commodities in order to bring their prices in line with those prevailing in the world market. This option, although attractive, faces the problem of uncertainty concerning its effect on the trade of foodstuffs. The export of food products was shown in our results to be the major source of benefit from the CAP and, moreover, it was almost exclusively of an intra-union nature. This being so it seems that the risks of damaging exports are very much reduced and might even be over-compensated for by consumption effects. However, successful implementation of this scheme would require that production and supply were not seriously affected, which could be assured to some extent by protection against import penetration from outside the union. An intermediate option which includes the reduction of protected prices designed to favour those crops where Europe has a comparative advantage, while at the same time considering the need for a more equal sharing of the benefits, seems to be the most promising substitute for the radical option of a complete scrapping of the scheme.

Turning now to factor mobility the results have been very meagre given that this was one of the main objectives of the Rome Treaty. It was seen that labour mobility was almost insignificant and in the case of Italy, the only country where it was possible to estimate its effects on growth, there was only a very small gain on this account.

The effects of direct foreign investment are more visible although without a clear net outcome. In fact, during the nineteen-sixties its impact in terms of growth was negative but in the following period this result was generally reversed. As the estimates for this period are more reliable one can concede that in general integration was beneficial on this account, although only in the cases of Denmark, Germany and Ireland were the gains significant. Furthermore, we can confirm the so-called tariff discrimination hypothesis and that some community countries have attracted substantial American investment (namely the U.K., Germany, Ireland, Denmark and Italy) which, nevertheless, has been counter-balanced by an opposite flow across the Atlantic, mostly by British firms.

Another expected outcome of economic integration derives from the possibility of improving the terms of trade. In this regard, and following Petith's (1977) work, the belief has been widespread [see Jones and El-Agraa (1981), Mayes(1983) and Robson (1984)] that these gains could be the most important source of benefit from integration within the EEC. This is no doubt so because until now[6] there have been no other empirical investigations of this question. However, our results do not confirm this optimism. In fact, these effects have not been the single most important source of gain and for the 1960's we even obtained a generalized loss on this account. Nevertheless, given the theoretical and empirical difficulties involved in this issue, we have some reservation concerning our own results. We would prefer to consider this issue an "in dubio" situation, which reinforces the need for a much deeper investigation in this area covering both the theoretical and empirical sides of the question. However, we did confirm some of the theoretical predictions about relative gains, in particular those (eg. Robson 1984) which state that the larger the economic area the greater is likely to be the improvement in its terms of trade.

Finally, from now on it is obvious that the terms of trade issue within the EEC will not be exclusively determined by tariff considerations. In fact, it is likely that the operation of the European Monetary System will affect the terms of trade more significantly than tariff considerations. In fact, although we have not been able to consider the EMS because it has been in operation only during the last three years of the period covered by our study, we are aware that the incorporation of foreign currency considerations in our model is likely to be the most needed and fruitful approach for any future developments of the balance of payments alternative framework to customs union theory that we have been presenting.

We must now outline some broad strategic considerations and policy issues that, in our view, confront the European Community today.

The major challenge that the EEC will have to face if it wants to continue to reap benefits from integration it is the renewal of old and the discovery of new ways to expand foreign trade, which has been shown to be the major source of growth and which has been shown to be facing possible exhaustion. In our view this can be achieved through the simultaneous use of two basic policies, one targeted to further enlargements to include other European countries and the other related to the implementation of an active policy of export promotion at the Community level.

Indeed, they are closely interrelated as is well illustrated in the current process of Mediterranean enlargement. In fact, if we are to rely on the past experience of other members (mostly Italy and Ireland) it is not difficult to foresee that the new members might be strongly and negatively affected by a deterioration of their terms of trade and an increased propensity to import. On the other hand, and contrary to what has happened in the case of Ireland, it is not foreseeable that either Spain or Portugal can significantly expand their exports to the EEC given that both already have Association Treaties that guarantee them relatively easy access to the Community market. In the case of Portugal, these prospects might be still further aggravated by the fact that under the present system of financing the CAP it is likely that after the transition period it would become a net contributor to the European budget.

If the EEC is going to prevent the further divergence of an already divergent Community and to provide the gains from integration in Europe which new members will undoubtedly anticipate, she will be faced with an ever growing demand for increased help through the Common policies. However, we saw that in this regard the CAP is perverse in the sense that it increases inequality between countries and that the other common policies lack sufficient scope to have a major counteracting impact on growth. Furthermore, if these common policies were going to be increased in order to achieve this goal the Community would, very likely, face an unbearable expansion of the present budgetary problems.

We would, therefore, advocate an alternative policy to promote exports (mostly of industrial products) from the peripheral countries as the main instrument to stop divergence within the Community. Several mechanisms can be created to this end and we would favour as the core of the scheme the establishment of a special fund to finance imports into the Community that originate in those members, and the working of a dual exchange rate system (say a "blue rate" to have a parallel with the green rates) also targeted to that objective. Further, other short range policies such as an immediate lifting of all quantitative restrictions that the peripheral countries still face and some promotion of foreign investment in the south, could also be very helpful.

Regarding further enlargements the EFTA members should be the next to join the Community, in particular the Scandinavian countries, which up until now have had a severe burden resulting from their integration within EFTA (see next chapter) and could possibly benefit from full-membership status. This strategy for trade expansion should also include the implementation of special trade agreements with the New Industrialized Countries. In particular it should be remembered that Portugal and Spain already have special links with those in Latin America (i.e., Brazil, Mexico, Argentina and Venezuela) which would facilitate any future developments in this direction.

Last, but not least, there is the need for a global EEC strategy aimed at increasing exports of commodities and services with a high income elasticity of demand. In this respect the fundamental issues of the bilateral economic relations with Japan and the USA and the race for the new technologies must be considered. This last question should in fact be at the centre of new and reformulated Common policies, but the Community should avoid the risk of embarking upon a blind race for new technologies. The objective here should clearly be the enhancement of the content of European exports.

In relation to bilateral trade with Japan and the USA we find (see Chapter V) that although there was some trade diversion in relation to the United States the Japanese have achieved an enormous import penetration in European markets for the past twenty-five years, a fact which justifies the use of transitional protective measures by the Community. As far as the USA are concerned the main problems are centered upon the capital and foreign exchange markets which, under the present circumstances, should not be separated from any trade arrangements.

1. That is, assuming the same pre-union tariff and that price elasticities and cross elasticities are not sensitive to price changes which, as he recognises, are awkward assumptions.
2. See beginning of this section and expression 3.49 in chapter III for a list of what might come under this heading.
3. Its value should then be seen jointly with the error term and the interdependence term.
4. For reasons of uncertainty in the estimates for France we took only 1/3 of her loss in terms of GDP growth rate (see also section 7.4). If the full loss was taken into account then the result would be reversed and the EEC would have registered a loss.
5. Even admitting that the trade balance loss was overestimated due to difficulties in estimating the trend growth rate of this variable.
6. Apart from some results of general equilibrium models, namely by Miller and Spencer (1977) and Viaene (1982).

Chapter Seven

EFTA AND GROWTH IN EUROPE

7.1 - The Free Trade Area in the sixties

We now look at the performance of EFTA countries. We must start by adding some extra qualifications to those already referred to in Chapter V, concerning the reliability of the trade effects estimated. First, it was not possible to include the home market (except in the case of the U.K.) and therefore both the extra - and intra - area trade effects will be biased. The sign and magnitude of such bias will depend upon the relative changes of both home and external markets (according with the expression presented in footnote 1 of Chapter V). But, if one admits that the effects of changes in the home market are positive and much smaller than those of the trade flows then the estimated trade effects will be downward biased. Second, the fact that we do not have the values for trade between EFTA and non-member countries, with the exception of the EEC members, means that we have aggregated all the other countries into the rest of the world which makes the aggregation bias more serious. The direction of bias cannot be foreseen, and it might be quite substantial (see Chapter V). Furthermore, in the case of EFTA, the possibility of checking our estimates of the integration effects on trade against those of other studies is more limited, since their number is rather small. On the other hand, the major studies available (EFTA 1969 and 1972) also used a share approach which makes them particularly suitable for comparison. There are, however, a few points that must be made prior to any comparison of results. These include the use of a different data base, which in the EFTA case was disaggregated at the commodity level and covered only manufactured goods while ours is for total trade, and uses different levels of aggregation at the country level and a different base year (1958-59-60 in our case and 1959 in the EFTA study). The results for comparison are presented in table 7.1, below.

The striking feature is that the intra-area effects are basically similar[1], while the extra-area effects are completely different. The EFTA study recognizes that their estimates might be overestimated. The reasons given for this are an underlying trade creation process

Table 7.1 - Comparison of *own*[1] estimates with those of EFTA for 1967
million US dollars at 1967 prices

EFTA COUNTRIES	Intra-area effects				Extra-Area effects Trade *diversion*[2]					
	IMPORTS		EXPORTS		EEC		USA		ROW	
	Own	EFTA	Own	EFTA	Own	EFTA	Own	EFTA	Own	EFTA
Austria	133	151	252	148	-223	46	..	7	-317	9
Denmark	139	437	99	174	-252	59	-22	20	105	99
Finland	49	128	60	216	-239	128	..	-10	-196	-23
Norway	135	309	125	167	-398	55	..	24	58	29
Portugal	33	32	124	111	-105	14	..	1	19	-1
Sweden	475	498	245	494	-691	181	..	68	-403	16
Switzerland	157	194	267	206	-785	57	..	-12	-386	2
United Kingdom	63	455	57	683	1514	38	218	-3	-5057	110

Source: EFTA (1972) - tables 9(c) and 10. Notes:

1 - Own estimates were converted into 1967 prices using the corresponding export and import implicit price deflators.

2 - Trade diversion is here taken as equivalent to our estimate of changes in total import flows from non-member countries.

generated by trade liberalization which was brought about by the GATT tariff negotiations; the EFTA origin system, which may have induced imports of intermediate products, and the adjustments due to the Common Customs Tariff of the EEC. These are not, however, sufficient to justify the disparity between the two estimates. Therefore, this large difference requires some explaining.

The main reason for differences in results is to be found in methodological differences. The estimation procedure used by EFTA is as follows:

$$I^{NP} = M_{67}^{NP} - [(n_{59} - n_{54})\frac{8}{5} + n_{59}]C_{67}$$

where,

I^{NP} = integration effect on imports from countries not belonging to EFTA

M_{67}^{NP} = imports in 1967 from countries not belonging to EFTA

C_{67} = apparent consumption in 1967 (domestic production + imports - exports)

$$n = \frac{M^{NP}}{C}$$

Table 7.2a - Total trade effects of integration 1961-66
million ECU's at 1975 prices

-		1961	1962	1963	1964	1965	1966
U.K.*	X	-218	-489	-1049	-1503	-2246	-3352
	M	-532	-1359	-1563	-2257	-3657	-5944
Denmark	X	-22	63	87	101	27	5
	M	63	60	150	172	231	183
Norway	X	-17	-15	6	51	102	164
	M	-114	-177	-237	-247	-276	-241
Sweden	X	19	49	87	129	155	209
	M	-176	-326	-428	-455	-566	-734
Finland	X	16	6	-13	-22	-29	-38
	M	-43	-231	-234	-246	-255	-435
Switzerland	X	15	64	70	85	114	197
	M	-155	-279	-390	-545	-738	-960
Austria	X	-37	-84	-136	-168	-203	-212
	M	-87	-157	-212	-244	-281	-377
Portugal	X	0.1	7	31	56	85	102
	M	-5	-7	-38	-29	-19	-65

It can be seen that the method uses a gross estimate of the trend in the change in the share of imports, which is based on two point estimates and a linear extrapolation. More significant is the fact that the period 1954-59 is used as a basis for extrapolation, yet 1954 is a year that was greatly influenced by European reconstruction, the Marshall plan and the Korean war. This would explain a decrease in the share of imports and consequently an estimated external trade creation effect.

Finally, reference should be made to the case of the U.K. which, in our estimates, and unlike other countries, showed substantial

Table 7.2b - Total trade effects of integration 1967-72
million ECU's at 1975 prices

-		1967	1968	1969	1970	1971	1972
U.K.*	X	-4934	-6397	-7625	-8161	-9361	-10858
	M	-9127	-12350	-16148	-18756	-21540	-24348
Denmark	X	-81	-174	-306	-424	-494	-474
	M	4	-137	-278	-430	-689	-814
Norway	X	243	314	324	274	306	414
	M	-291	-393	-504	-595	-740	-892
Sweden	X	307	412	520	561	645	746
	M	-954	-1091	-1155	-1469	-1931	-2580
Finland	X	-63	-59	-38	-65	-62	-116
	M	-847	-1188	-1367	-1408	-1579	-1863
Switzerland	X	342	451	468	414	433	544
	M	-1145	-1292	-1371	-1426	-1542	-1626
Austria	X	-228	-221	-236	-260	-242	-264
	M	-508	-666	-721	-740	-676	-641
Portugal	X	102	112	109	98	107	99
	M	-75	-59	10	62	68	80

* Values for 1967 and following years are unreliable due to a large devaluation (14.3%) of the pound sterling.

external trade creation with the EEC. This is not simply due to the fact that only in the U.K. case could we take account of the home market. Even without the introduction of the home market the value would still be positive; although only around 26% of the estimate presented in table 7.1. Moreover, these estimates were also greatly affected by shifts in trade with the former colonies, as in the case of France.

However, one can not be sure of having a similar outcome for the estimates of the other countries because this depends on satisfying the condition set out in footnote 1 of chapter V. Nevertheless, we think that in general these estimates would not be significantly affected and can be considered reliable. So, we can proceed by examining the results presented in table 7.2, above.

The analysis of the weighted share indices clearly showed (see Mendes 1985a) that all members had a significant increase in their intra-area trade. On the other hand, trade with the EEC was also very much reduced (with the exception of the U.K.), particularly in the case of the Scandinavian countries. As far as the particular result obtained for the U.K. is concerned it raises several questions. To begin with, it is very doubtful that it can be attributed to trade deflection resulting from tariff differentials because her tariffs were higher than those of the majority of other partners. However, if one considers the size of her domestic market relatively to the other partners it is not difficult to conceive that the U.K. could benefit from production and investment deflection. In this regard both the large gains obtained in terms of gross foreign investment (see below) and the fact that the United States also increased their exports to the U.K. seem to reinforce this hypothesis. On the other hand, the fact that the Commonwealth countries lost their relative preferences in the U.K. market explains, to some extent, the huge reduction of imports from the rest of the world. As far as the other countries are concerned, it is more difficult to admit that any significant trade deflection might have occurred, except in the cases of Norway and Denmark, which being low tariff countries experienced an increase in trade with the rest of the world (see table 7.1). However, a similar result for Portugal is more likely to suggest that production and investment deflection has arisen due to lower wage costs.

Nevertheless, trade reduction with the EEC and the rest of the world was generally so large that all countries show a reduction of total imports (see table 7.2 above). This is a very striking result which contrasts with a priori expectations and the EEC's own experience referred to above. In fact, this means that trade creation was most unlikely (with the exceptions of Denmark and Portugal) or, if it existed, it was so small that it was completely offset by trade suppression. Moreover, the fact that exports only showed a small increase (or even decreased as in the cases of U.K., Finland and Austria) reinforces this conclusion unless one is prepared to accept that there was a large suppression of intra-union trade. However, one can not say that this suppression of trade has worsened the trade balance because all but Denmark registered some improvement.

Nevertheless, the basic conclusion[2] is that for this period, and contrary to widespread belief, the European Free Trade Association is very far from being a successful story of trade promotion during the 1960's.

The overall outcome is a negative effect on the GDP growth rate achieved by all the member countries except Portugal. The results provided by the balance of payments framework are shown in table 7.3 below.

The loss in the case of the U.K., Denmark and Finland is very substantial. The sensitivity tests performed confirm this result.

Turning now to the various growth components we see that, in general, the terms of trade played a minor role[3]. However, they generally had a positive contribution to growth. The EFTA countries also benefited from a decrease in their propensity to import. This is contrary to what one would expect from the formation of a free trade area, given that trade liberalization would increase both exports and imports. However, given the circumstances that surrounded the formation of EFTA, it is understandable that their members have reacted by adopting a more protectionist attitude towards non-

Table 7.3 - Integration effects on the % trend growth rate of member countries
- EFTA 1961-72 -

Contributions to growth	United Kingdom	Denmark	Norway	Sweden
Actual growth rate	2.75	4.13	4.08	3.94
1 - Growth rate due to EFTA [1 = -2-3+4+5+6+7]	-3.69	-2.65	-0.42	-0.78
Which was made up of				
2-Terms of trade changes	-0.05	0.90	0.07	-0.14
3-Change in propensity to import	-2.38	-0.15	-0.13	-0.46
4-Growth of export volume	-3.63	-1.58	1.34	0.83
5-Change in the trade balance position	-3.62	-0.94	-1.49	-2.22
6-Foreign investment	1.02	0.06	0.12	-0.01
7-Residual + errors [7 = (1+2+3)-(4+5+6)]	0.11	0.56	-0.45	0.03

Table 7.3 (cont.) - Integration effects on the % trend growth rate of member countries
- EFTA 1961-72 -

Contributions to growth	Finland	Switzerland	Austria	Portugal
Actual growth rate	4.39	4.05	4.76	6.31
1 - Growth rate due to EFTA [1 = -2-3+4+5+6+7]	-2.59	-0.61	-0.40	0.16
Which was made up of				
2-Terms of trade changes	-0.38	-0.07	-0.03	-0.02
3-Change in propensity to import	-1.31	-0.59	-0.29	0.07
4-Growth of export volume	-0.32	0.60	-0.18	0.94
5-Change in the trade balance position	-3.63	-1.75	-0.62	0.04
6-Foreign investment	-0.28	0.004	0.01	-0.03
7-Residual + errors [7 = (1+2+3)-(4+5+6)]	-0.05	-0.12	0.08	-0.74

Notes: As defined in the model a negative sign of the terms of trade means an improvement;
-The total effect may differ from the sum of its components due to rounding.

members, particularly in relation to the EEC.

The performance in terms of export growth was also very poor. In fact, countries with a bad export performance (U.K., Denmark, Finland and Austria) experienced large losses on this account. Where gains were registered they could not even cover the offsetting effects of the changes in the trade balance. The only exception was Portugal, which curiously was the only country to benefit from a special protection scheme (the so-called Annex G) included in the Stockholm Convention.

Finally, in spite of EFTA (a free trade area) not being specifically targeted to promote factor mobility commercial policy has probably affected foreign investment. The evidence on this matter is limited for reasons given in the previous chapter, but our results suggest that these effects were small (and generally negative). The exception is the United Kingdom which experienced a gain due to large foreign investment inflows.

7.2 - An enlarged European Free Trade Area

As a complement to the enlargement of the EEC to embrace two former EFTA members (the U.K. and Denmark) the two organizations negotiated the creation of a European free trade area between their members, covering most industrial products and with a few provisions for foodstuffs, which was in full operation by 1977. This is a very interesting situation because it covers two alternative options concerning the choice of forming a customs union or a free trade area. This has been the subject of theoretical dispute which we will examine in more detail in the next section. However, it is convenient to remember that in the case of free trade areas the so-called trade deflection effects are likely to be verified unless very tight "origin rules" are implemented to prevent third countries switching their exports towards those members with lower tariff rates. Furthermore, even when there are efficient "origin rules", it is possible to have indirect trade deflection of final products (see Robson 1984). Under these circumstances it is possible to foresee a situation in which countries reduce imports from partners while increasing them from non-member countries. In fact, the EFTA experience seems to confirm this possibility. That is, EFTA countries reduced imports from their former colleagues while increasing them from the United States (with the exception of Portugal which benefited from special protection). Moreover, most of them also registered a decrease of imports from the new EEC partners in the wider free trade area (with the exception of Austria and Portugal). All the former EEC members registered a significant increase in their imports from EFTA. In particular it is interesting to contrast the EFTA values[4] with those of the former members (U.K. and Denmark) which registered a reduction of imports from the United States and an increase in imports from the EEC.

Despite the fact that some of the EFTA countries had external tariffs lower than the Common Customs Tariff and the fact that EFTA has adopted the much stricter "origin rules" imposed by the EEC, these were not sufficient to prevent American exporters from using the EFTA countries to penetrate the EEC market[5].

However, the most striking feature of the results is that imports

Table 7.4 - Total trade effects of integration for manufactured goods (1974-81)

million ECU's constant prices

years		1974	1975	1976	1977	1978	1979	1980	1981
Austria	X	171	382	322	445	602	782	800*	800*
	M	46	253	530	790	994	1109	1028	848
Finland	X	-30	41	125*	125*	125*	182	691	1077
	M	119	123	-85	-353	-458	-333	-267	-258
Iceland	X	-4	-3	-3	-0.4	-0.9	-0.2	-0.2	-0.2
	M	-0.1	46	25	17	-20	-14	-8	-3
Norway	X	32	-50	-314	-513	-715	-750	-816	-854
	M	154	329	292	29	-488	-701	-807	-862
Portugal	X	-46	-150	-330	-310	-116	-40	-29	-109
	M	-127	-221	-305	-303	-325	-284	-213	-160
Sweden	X	204	276	-248	-624	-920	-909	-1064	-1256
	M	243	424	-489	-1010	-1340	-914	-1395	-1874
Switzerland	X	54	254	395	810	745	607	-45	-432
	M	-350	-766	-1223	-863	-718	-22	-1037	-1903

* Corrected for unexplained variations of trade in transport equipment (product 7).

from America were not sufficient to offset the reductions of imports from within the wider free trade area (EFTA-EEC) and some other countries. In fact, it can be observed from table 7.4 above that, with the exception of Austria, all the other countries experienced a reduction of imports. Once more, the basic explanation might lie in a process of production and investment deflection, although this time trade deflection also has played a significant role. This hypothesis is justified on the grounds that imports of raw materials and intermediate goods from the EEC showed some increase (broadly products 2 and 5).

Therefore, as in the nineteen-sixties, it is most unlikely that there has been any net trade creation[6] during the period under discussion, rather it seems likely that trade suppression might have been very significant. This result is further reinforced by the fact that most countries also had a poor export performance (only Austria, Finland and Switzerland show induced export growth).

Table 7.5 - Integration effects on the % trend growth rate of member countries
- EFTA 1974-81 -

Contributions to growth	Austria	Finland	Iceland	Norway
Actual growth rate	2.76	2.83	3.69	4.25
1 - Growth rate due to EFTA [1 = -2-3+4+5+6+7] Which was made up of	1.02	-0.61	-0.91	-4.01
2-Terms of trade changes	0.37	-1.08	-0.53	-0.575*
3-Change in propensity to import	0.72	-0.28	-0.23	0.03
4-Growth of export volume	0.71	4.05	0.17	-2.84
5-Change in the trade balance position	1.69	-2.69	-1.44	-1.72
6-Foreign investment	0.15	-0.18	0.09	0.01
7-Residual + errors [7 = (1+2+3)-(4+5+6)]	-0.43	-3.14	-0.49	0*

Table 7.5 (cont.) - Integration effects on the % trend growth rate of member countries
- EFTA 1974-81 -

Contributions to growth	Portugal	Sweden	Switzerland
Actual growth rate	3.94	1.21	0.92
1 - Growth rate due to EFTA [1 = -2-3+4+5+6+7] Which was made up of	0.08	-2.59	-0.62
2-Terms of trade changes	0.03	-0.406**	1.33
3-Change in propensity to import	0.02	-1.83	0.85
4-Growth of export volume	0.47	-3.12	-2.77
5-Change in the trade balance position	-0.88	-1.11	0.02
6-Foreign investment	-0.12	-0.44	-0.10
7-Residual + errors [7 = (1+2+3)-(4+5+6)]	0.66	-0.167	4.41

Notes: Such as defined in the model a negative sign of the terms of trade means an improvement; The total effect may differ from the sum of its components due to rounding.

* Estimated jointly with the residual. ** Estimated without including total capital flows in expression 3.41).

The overall impact shows a negative effect on the growth rate of output, with the exceptions of Austria and Portugal. All the other countries suffered quite large losses[7], some in excess of the actual growth rate achieved. Once more the Scandinavian countries were the most affected.

Turning now to the various growth components we see that the terms of trade showed a general positive contribution to growth as in the nineteen-sixties, although with two exceptions (Austria and Switzerland). However, the values are very unstable, according with the corresponding tests of sensitivity[8] In fact, any of these countries has a probability of having benefited from terms of trade of at least 30%.

With regard to the propensity to import, most countries had an increase, with the exception of Sweden, although without very large losses in terms of GDP growth. The two higher values (for Austria and Switzerland) are not reliable, but it may be noted that during this period Switzerland was the only country to have experienced a positive shift in the income elasticity of demand for imports.

Four countries show gains in terms of export growth; however, part of the gains was offset by a negative effect from changes in the trade balance. The most notable exception is Austria. Also, in terms of foreign investment, this is the only country to have benefited significantly from factor mobility[9]. So, once more, we see that the effects of integration were rather small and may even be negative[10], though we must remember that these are only gross estimates.

7.3 - Free trade areas vs Customs unions - An EEC/EFTA comparison

As stated earlier, it is important at this stage to address the question of knowing whether customs unions are preferable to free trade areas. It is clear that in the case of the European experience customs union is certainly preferable. In fact, not only did it increased the income of their members, which must be matched against a negative effect for the EFTA, but also those members (U.K. and Denmark) that moved from EFTA to the EEC showed an improvement in terms of integration induced growth.

However, this result needs further scrutiny in order to assess whether this experience can be applied to other circumstances; to see if it matches theoretical predictions, and to identify the reasons for such disparate performances. Beginning with a brief examination of the theoretic contributions on the issue [eg. Balassa (1961), Shibata (1967), Curzot (1974), El-Agraa (1981) and Robson (1984)] we face the rather curious situation that although some authors spend several pages showing that free trade areas are superior to customs union they usually conclude by conceding that customs union are in fact preferable to free trade areas. The basic reasons are fivefold:

(a) the assumption that an efficient origin system can be implemented is very questionable;

(b) customs unions are not tariff-averaging as is usually assumed;

(c) the EEC and EFTA correspond to one of the specific situations where the supremacy of a free trade area is not verified;

(d) the assumed complementarity between economies required to avoid substantial production and investment deflection did not occur;

(e) free trade areas incur substantial administrative costs to maintain their individual tariff systems and to implement the "origin system".

However, before assessing the likelihood of these situations relatively to the EEC and EFTA we shall briefly consider the theoretical situation which is often used to illustrate advantages of the free trade areas.

The theory usually assumes that in the pre-union situation both countries had prohibitive tariffs but that one of them (say the partner) was more efficient than the other (the home country). After establishing the free trade area the equilibrium price in the home country will be lower than the one brought about by a tariff-averaging common customs tariff, while the price in the partner country will remain at the previous level. In these circumstances the partner producers would shift their production to the home country leaving the less profitable domestic market to the rest of the world. Such an outcome would certainly be superior to a customs union, not only because trade creation in the home country would be larger, but also because the substitution of domestic production in the partner market for that of the "rest of the world" would bring an extra gain through tariff revenue collected by the government.

However, one can easily argue that the occurrence of such a case is most unlikely. To begin with, it is unlikely that the equilibrium price for the home country in the post-union situation would be lower than the equilibrium price with a common customs tariff and it does not seem plausible that the most efficient of the members (the partner) has necessarily a more elastic demand curve. In fact, we can foresee a situation in which the partner country producers begin by pricing their goods in the home country market at a level just below the rest of the world (tariff inclusive) price, while at the same time slightly increasing the price in their domestic market. On the other hand, the "rest of the world" producers, faced with displacement in the home country market, can be imagined to react by reducing their prices in this market (it must be remembered that they are assumed to be more efficient) while increasing their prices in the partner country so that the resulting terms of trade gain might offset the losses in the home country market[11]. It is then possible, for given slopes of the demand and supply schedules, to end up with an equilibrium price that is higher than the one arising from a (tariff-averaging) customs union and even with a decrease in total imports if the size of the partner's market is much larger than that of the home country.

Now, apart from this more realistic possibility, let us see if in the EEC-EFTA case there are reasons to believe that some of the circumstances enumerated above can be met. Excluding the factor of higher administrative costs, since it has little relevance, we shall move to the third possibility, which occurs in a case in which both partners are relatively inefficient and the demand in the home market is either very large or very elastic. This has been shown (Curzot 1974) as being one case where customs union are preferable to a free trade area. Using this definition we see that for the period 1974-81, with regard to a large number of products, the EEC might represent the home country and EFTA the partner country. Further, there is no reason to suppose that the Community's common customs tariff might be equal to or greater than the pre-integration average tariff (case b).

Turning to the question of the "origin system", the main problems raised are concerned with assessing whether such a system can work effectively. There are fears that these rules might be either too liberal, and therefore cause large trade deflection, or too strict, causing a severe reduction in trade. Because of this the interpretation of our estimates of trade effects is rather complex. They do, in fact, suggest that there had been some trade deflection which was higher in the 1970's, despite the fact that during this period origin rules were stricter. However, this contradiction is more apparent than real given that in the EEC-EFTA free trade area there are more competitive situations (and non-complementarity) than there were in the initial free trade area. Further, the strictness of the origin system adopted in the 1970's, coupled with increased protectionism, combine to justify the trade reduction.

However, the fact that during the 1960's there was also a reduction in trade cannot be explained on these grounds. A more likely explanation is to be found in production and investment deflection effects. In fact, it must be remembered that these may occur regardless of an efficient origin system. Moreover, the EEC common customs tariff, when applied to intermediate products, is certainly higher than the average tariff duties applied by EFTA to the same goods. This fact, coupled with a likely domestic cost advantage, leads to large production and investment deflection.

We can then conclude that not only does the supposed superiority of the free trade areas have a weak theoretical backing, but also that in the particular EEC-EFTA case we are likely to find most of the conditions generally accepted to make customs union more advantageous. Our final results in terms of GDP growth rates clearly confirm this hypothesis.

However, it is still worthwhile to examine a comparison of the various effects that contributed to this result. Starting with the trade effects, we see that the differences between the two unions are very large in both periods: while the EEC experience in the two periods shows large increases in both exports and imports, EFTA generally suffered trade reductions. Obviously, in terms of growth effects, these differences were partly reduced due to offsetting effects through changes in the propensity to import.

As far as the terms of trade are concerned the comparison is more complex. While in the nineteen-sixties EFTA clearly outperformed the EEC, during the nineteen-seventies this advantage was not so obvious. It must be pointed out that these results are in opposition to a widespread belief that customs union are more likely to improve the terms of trade of their members. Despite the fact that our estimates are not exempt from bias for reasons given in previous chapters we do not see strong reasons to dismiss this result. First of all, contrary to the theoretical predictions, the free trade area was more trade diverting than the customs union. Furthermore, if two different equilibrium prices are to co-exist in each member country in the manner we described above, there are reasons to expect that the free trade area will improve her terms of trade. Also, the fact that EFTA is made up of smaller and lower tariff partners is likely to result in a better performance in the terms of trade. Finally, the closing of the gap verified in the enlarged free trade area is also consistent with the prediction that the larger the union the more likely it is to obtain terms of trade improvements.

Comparing the relative performances in terms of factor mobility (foreign investment) we also did not see any marked superiority in the customs union, in spite of there having been special policies targeted to this objective. The results are very mixed and for the 1970's they are not strictly comparable[12]. However, the broad conclusion is that in both cases they were rather small. In relation to a better performance of EFTA during the nineteen-sixties, a plausible explanation may be the existence of significant investment deflection effects.

In fact, this possibility reinforces the final conclusion that the distinct performances of EEC and EFTA have to be explained mainly through the deflection of production and investment brought about by tariff changes and the operation of "origin systems". However, it is obvious that our results are only a first indication in this direction and that they need to be checked against further research at the theoretical level and against direct estimates of these effects.

Nevertheless, our conclusion still is that in the European context customs unions are preferable to free trade areas.

7.4 - Integration, growth and cumulative causation in Europe

We are now in a position to conclude with an evaluation of the European integration process at a global level. In the light of our results we can provide a quantitative assessment of the impact of integration upon income: namely, that for the first 25 years of integration, Europe (EEC+EFTA) has benefited from a gain equal to at least 1.5% of her GDP in 1981. There were, however, inter-country differences and there was a big disparity between the two trading blocks. While EFTA experienced substantial losses estimated to amount to around 11.5% and 6.6% of her GDP in 1972 and 1981 respectively, the EEC had an estimated gain around 2.2% and 5.9% of her GDP in those years[13]. Furthermore, the 1981 value for the EEC would have been 1.7 percentage points higher if it were not for the operation of the Common Agricultural Policy. In total, Europe had a loss of 2.5% of her GDP in 1972 and a gain of 3.9% in 1981, which makes up the 1.5% net gain for the full period of integration.

We can also give an answer to the question of whether integration has increased or decreased the inequality between countries. Using the estimates of GDP with and without integration we can compute Atkinson's measure of inequality of per capita income between member countries and its changes are presented in table 7.6 below.

Table 7.6 - Effects of integration on inequality in per capita income between countries

Indicators	% change in Atkinson's coefficient		
	$\epsilon=1$	$\epsilon=1.5$	$\epsilon=2$
Period 1961-72			
EFTA	-6.67	-11.19	-13.07
EEC	-32.14	-31.03	-30.51
EUROPE (EFTA+EEC)	-16.42	-18.69	-17.01
Period 1974-81			
EFTA	-13.33	-11.88	-10.23
EEC			
(1) with CAP	15.38	13.56	12.66
(2) without CAP	-2.56	-1.69	-1.27
(3) inequality increase due to CAP [3 = 1+2]	17.94	15.25	13.93
EUROPE (EFTA+CEE)	-1.64	0	1.60

Source: Mendes (1985a).

Broadly speaking we can then say that integration has not only increased income but has also increased equality[14] between the participating countries. However, a detailed analysis of the results raises some important questions.

First, EFTA made an important contribution to this result but, since EFTA had a negative effect on growth, one can say that this was partly an immisering reduction of inequality.

Second, during the 1970's the EEC faced a divergence process, for which CAP appears mainly responsible. However, even without CAP, the EEC would have made only a minor contribution to equality, which contrasts with her results for the nineteen-sixties and the EFTA performance. This gives some support to the "cumulative causation" thesis that the larger and more heterogeneous is the area the more likely it is to generate inequality.

So, even if CAP were scrapped, it is hardly conceivable that the EEC could go on generating equality between its members after the enlargement to the Mediterranean periphery. This prediction can be further justified by two other indicators. One is that when we increase the inequality aversion factor (ϵ), that is, when we give more weight to those countries at the bottom of the income scale, then the inequality reduction is less pronounced. The other is that all lower income countries (ie, Ireland, Italy and U.K.) were in general the ones with the smallest gains.

Furthermore, with the enlargement to the Mediterranean countries we have to admit as plausible that export induced growth is likely to be small due to the fact that those countries already had preferential regimes, and taking into account the contributions from the other various mechanisms (namely terms of trade and factor mobility) one cannot be optimistic about the possibility of a reduction in divergence. Also, common policies are either not sufficiently developed to counteract this trend or, as is the case with the CAP, they may even reinforce divergence. It is in fact important to note that CAP generated large inequality, in spite of having benefited the Irish economy, which effect is clearly reflected through the decreasing values when inequality aversion (ϵ) is higher. So, a challenging task that an enlarged and more integrated Europe will have to face will be the development of adequate policies to counteract this trend towards divergence of per capita incomes.

However, it must be stressed as a final conclusion that although the trend for divergence it is not very strong it is mostly due to differential trade performances. This being the case, any trade policies designed to enhance exports should necessarily allow for distributional considerations.

1. The fact that the intra-EFTA estimates are usually overestimated relative to ours (particularly in the cases of U.K., Denmark, Norway and Finland) is possibly due to the fact that the EFTA study decided to exclude any negative trade effects measured.

2. Even admitting that the reduction of imports is overestimated due to reasons presented in the beginning of this section.

3. The only exception is Denmark, which seems to have experienced a large loss, but this is uncertain given that for a 75% confidence interval the values vary between -0.1 and 2 percentage points.

4. However, it should be remembered that in the case of EFTA it was not possible to include the home market so that their values may be underestimated.

5. In this case one has further reason to consider as totally irrelevant any supposed measures of trade creation/diversion by which to assess the welfare gains/losses from EEC membership.

6. It is interesting to note that only in the trade of food products was there some increase of imports.

7. The loss for Norway is clearly overestimated due to changes in trade of chemicals and transport equipment that are not due to integration but, possibly, to North Sea oil related investments.

8. Here, the results concerning Norway were based on an estimation with total capital flows in expression 3.41 which produced values quite different from the ones presented in table 7.5 above.

9. This in itself raises some doubts as to whether these benefits can be entirely ascribed to integration.

10. The most striking result in this regard is the case of Portugal which being the poorest country did not manage to attract integration induced investment in spite of the exceptional setting up by EFTA of a Fund for the Industrial Development of Portugal. However, it must be pointed out that it was not possible to provide for the explicit inclusion of this Fund (whose donation component was rather small anyway) and that during this period Portugal went through substantial political and social unrest that certainly affected these estimates.

11. Alternatively, or in conjunct, the partner country can be imagined to increase its tariff rates in order to keep his position in the domestic market.

12. It should be remembered that due to data difficulties we could not obtain direct estimates of foreign investment for EFTA what makes them conceptually different (see chapter III).

13. Due to possible overestimation of the results for France and the United Kingdom during the sixties (due to the effects of trade with the former colonies) and for Norway in the 1970's (due to North Sea oil) we used only 1/3 of the impact estimated for the growth rate.

14. It is important to note that if as claimed [eg. Kiljunen (1980); Molle et al (1980), and Keeble et al (1982)] the growth of GDP is the most important effect in explaining regional performances then it is likely that integration has also reduced regional disparities.

Chapter Eight

CONCLUSIONS

We have dealt with the development of both an alternative framework to conventional Customs Union approach and uses of this new framework to measure the effects of the European experience of economic integration; and although we have been presenting the various policy and research issues throughout the text it is now useful to summarize the main points. We shall start with the theoretical contributions, remaining insufficiencies and some new directions for future research.

We have added further reasons for the widespread discontent concerning the assumptions of customs union theory and its exclusive reliance on the concepts of trade creation and trade diversion. It was shown that the assumption of an automatic adjustment in the balance of payments at the union level would require that there is no trade diversion and that at the country level the further requirement exists that all countries share equally any trade creation gains. Such a condition could only exist in the unlikely case of two identical partner economies. However, our empirical research has shown substantial changes in the balance of payments, which makes this assumption of customs union unacceptable. It was also shown that ex-post measures of changes in trade flows will include various effects and not just those of trade creation and trade diversion, namely: trade reorientation, external trade creation and trade suppression. Therefore, the closest measurement of trade creation will be the effect of integration on total imports (rather than on intra-union imports) but the requirement that trade reorientation and trade suppression offset each other still exists if the total trade effect is to be equal to trade creation. Furthermore, we have expanded on the suggestion that ex-post measures of trade creation/diversion effects should be interpreted as range values dependent on the assumptions made about the partners supply curve. In fact, given the level of aggregation at which empirical studies are always conducted we have to consider that the "normal case" will be the one in which, in the pre-integration stage, the partner country already has a significant share of the market. We have shown that in these circumstances the measurement of trade creation/diversion effects may be misleading or even impossible. We therefore conclude that the customs union approach to measuring integration effects is in a deadlock situation[1].

In the balance of payments framework which we have developed to measure the effects of integration we discarded the need to use the notions of trade creation/diversion. Instead, we used the effects on total trade which are related to output growth (this is a simpler and more meaningful concept than the welfare effects) through the use of the foreign trade multiplier. We have shown that this accounts for most of the growth rate of European economies and deviations from this long-run equilibrium growth rate brought about by capital flows, terms of trade changes and price volume effects were also estimated. The new framework allows us to examine the relationship between growth and the various components of trade and other integration effects, as these affect the balance of payments, therefore providing an integrated measurement of all widely accepted major effects of integration, that is: export growth, terms of trade changes, budgetary transfers and factor mobility (capital and labour). A fundamental novelty is that the new approach takes into account the effects on the import side operating through changes in the income elasticity of demand for imports. It is shown that the existence of a net trade creation effect is not a sufficient condition to obtain an increase in output, as this also depends on changes in the income elasticity. However, as it stands, the new framework is not itself devoid of shortcomings and requires further development. In particular, the measurement of the changes in the terms of trade is not satisfactory since we could not eliminate the interdependence term between these and the import growth rate. Moreover, the methodology can be substantially improved by including the effects of currency movements (particularly now that the EMS is in operation) and by taking into account the shrinkage of the home market that is bound to occur as a result of external trade expansion: these are the two most promising areas for further research.

Finally, in computing the overall effect upon the growth rate the new approach relies heavily upon the estimation of induced effects on total imports and on the income elasticity of demand for imports. Both present serious measurement difficulties. There has been no empirical work to date relating to changes in the income elasticity and we had to overcome this lack by using the dummy variable device. This is another area requiring further research. There are several techniques by which one can assess the trade effects and we chose the weighted share analysis. We have given a different presentation of this technique and have shown the importance of working at a very high level of disaggregation. We formulated some assumptions concerning the interdependence effect resulting from changes in trade flows between non-participating countries. We also introduced the home market which was seen to have a crucial effect on the estimates of the trade effects. Furthermore, we showed that a more realistic "anti-monde" is possible in which countries can be assumed to maintain constant relative growth rate differences so that any measured integration effects also remain constant. This extended version of the share technique was used to measure the effects of integration upon direct foreign investment flows.

Turning to the empirical study of the European experience of economic integration, we shall first consider an overall view of the integration effects. The main conclusion is that integration played a very important role in the growth of the European economies, which was estimated to be equal to at least 1.5% of their GDP in 1981. This value is higher than the 1% guess made in the late 1950's and thereafter widely adopted as the maximum benefit to be expected from integration. However, the gain obtained was not uniformly distributed throughout the period nor was it shared equally by all countries. The fact that the gain was obtained during the 1970's (a period marked by an upsurge of protectionism), raises the need for further research on the role of integration within a protectionist context. In relation to the two European trading blocks, EFTA and EEC, they were shown to have contrasting performances. While the EEC obtained significant gains during both periods, EFTA experienced large losses, which supports the view that a customs union is more advantageous than a free trade area. Once again further research is needed, this time to assess whether production and investment deflection effects are likely and can account entirely for the measured trade suppression. However, the disparity in the relative performance of the EEC and EFTA does not mean that integration has increased inequality in Europe. In fact, although inequality was not reduced during the 1970's (it was even slightly increased if we give a higher weight to inequality aversion) it had nevertheless been substantially reduced by integration throughout the 1960's. However, this result cannot be interpreted as refuting the circular and cumulative causation hypothesis that inequality is bound to be increased. First, in 1981 inequality was higher than it would have been without integration and with a high aversion to inequality. Second, much of the equalisation was obtained due to the EFTA losses (therefore, it can be described as an immisering equalisation). Moreover, the enlarged and more heterogeneous EEC experienced an increase in the inequality between their members. Although this is due to the operation of the Common Agricultural Policy the truth is that even without CAP it would have ceased to provide any contribution towards a reduction of inequality.

Turning to a more detailed examination of the results we find that integration has had a large effect on trade. In the case of the EEC there was a large increase in total imports which confirms the existence of substantial net trade creation effects. By contrast, EFTA has hardly showed any trade creation. Another important conclusion is that trade in foodstuffs has also been substantially affected by integration and should therefore be included in any measurement of integration effects on growth. It was also possible to estimate changes in the flow of direct foreign investment for the enlarged EEC and these confirmed the so-called "tariff discrimination hypothesis". As far as labour mobility is concerned the general conclusion is that it has not been very significant, but this result needs to be complemented by further research based on better data and improved modelling of these effects.

In relation to the various contributions to integration-induced growth we may conclude that so far factor mobility has played a minor role. We also have raised some reservations about the recently widespread and optimistic view that the terms of trade changes might have been the major source of gains. We have also examined the impact of the CAP in great detail and estimated that its cumulative inefficiency loss in 1981 was of around 1.7% of the Community's GDP in that year. Various alternatives to the scrapping of this common policy were considered and it was advocated that a less damaging policy would be one of selective price reductions which would allow for equality considerations and would be based on Europe's comparative advantage in the production of Mediterranean and temperate-zone food products.

The final, and in our view the most important conclusion is that most integration-induced benefits arise from export growth and to ensure that these do not reach exhaustion the European Community should adopt a global and strong policy of export promotion. Some of the items which we feel ought to be included in such policy are: (a) that importing of industrial products originating in the peripheral Community members should be encouraged by the use of import subsidies and/or a system of dual exchange rates; (b) that there should be further enlargement of the Community to include the remaining EFTA members, particularly the Scandinavian countries; (c) that special trade arrangements should be made with the so-called "new industrialized countries" aimed at an expansion of trade with these countries, and (d) that the technological policy should be directed towards an enhancement of the technological content of European exports and a shift of its composition towards products with a high elasticity of demand.

1. A review of attempts to include dynamic effects shows that these have been dealt with in a very crude and partial way.

REFERENCES

Abeele, Michel Vanden, Regional disequilibria, financial flows and budgetary policy in the European Community,
2 Economia, V(1), 117-153, Lisboa, 1981.

Aitken, N. D., The effect of the EEC and EFTA on European trade: A temporal cross-section analysis, American Economic Review, 881-892, December, 1973.

Armington, P. S., A theory of demand for products distinguished by place of production, IMF staff papers, 16, 159-176, 1969.

Atkinson, A. B., On the measurement of inequality,
2 Journal of Economic Theory, 2, 244-263, 1970.

Balassa, B, The theory of economic integration, Allen and Unwin, London, 1961.

Balassa, B., Trade creation and trade diversion in the European Common Market,
2 Economic Journal, LXXVII, 1-21, March, 1967.

Balassa, B., European Economic Integration, North Holland Publishing Company, 1975.

Balassa, B., Some effects of commercial policy on international trade, the location of production and factor movements, W. B. S. W. P., 236, June, 1976.

Barraclough, Geoffrey, The EEC and the world economy, 57-71, in: Seers et al. Integration and unequal development, MacMillan, London, 1980.

Begg, I., Cripps, F., Ward, T., The European Community: problems and prospects, Cambridge Economic Policy Review, 7 (2), 23-30, December, 1981.

Berends, Helena, Aspectos da Politica Agricola Comum da CEE, I. N. A., Lisboa, 1983.

Bieber, R., Jacque, J-P., Weiler, J.H.H. (eds), An ever closer union - A critical analysis of the Draft Treaty establishing the European Union, CEC - European Perspectives, Brussels, 1985.

Biehl, D., The impact of enlargement on regional development and regional policy in the European Community - in: Wallace-Herreman (eds) - A community of twelve: The impact of further enlargement on the EC, 211-246, De Tempel, Brugges, 1978.

Blancus, P., The CAP and the balance of payments of the EEC member countries, Banca Nazionale del Lavoro Quarterly Review, 127, 355-370, 1978.

Brown, A. J., Customs union versus economic separatism in developing countries, Yorkshire Bulletin of Economic and Social Research, 13, 33-40, 13 (1), 33-40, 13 (2), 88-96, 1961.

Burns, M. E., A note on the concept and measure of consumer's surplus, American Economic Review, 63, 335-344, 1973.

Camagni, Roberto, Cappelin, R., European Regional Growth and Policy Issues for the 1980's, Built Environment, 7 (3,4), l62-171, 1982.

Caves, Richard E., Intra-industry trade and market structure in the industrial countries, Oxford Economic Papers, 33 (2), 203-223, 1981.

CEPG - Cambridge Economic Policy Group, Policies of the EEC, Economic Policy Review (Chapter II), 1979.

Chenery, Hollis B., Ahluwalia, M.S., Carter, N., Growth and poverty in developing countries - in: H. Chenery (ed) - Structural change and development policy, Oxford University Press, 1979.

Collier, Paul, The welfare effects of customs union: an anatomy, Economic Journal, 84-95, March, 1979.

Cooper, C.A., Massel, B.F., A new look at Customs Union Theory, Economic Jornal, 742-747, December, 1965.

Corden, W. M., Economies of scale and customs union theory, Journal of Political Economy, 465-475, 1972.

Curzon, Victoria, The essentials of economic integration - lessons of EFTA experience, MacMillan, London, 1974.

Dayal, R., Trade creation and trade diversion: new concepts, new methods of measurement, Weltwirstschaftliches Archives, 125-168, 1977.

Denton, Geoffrey, Regional divergence and policy in the community with special reference to enlargment - in: Hodges - Wallace (eds) - Economic Divergence in the European Community, George Allen & Unwin - London, 135-149, 1982.

Dunning, J. H., Trade, location of economic activity and the multinational entreprise: a search for an ecletic approach - in : Ohlin, b. et al (eds) - The international allocation of economic activity, MacMillan, London, 1977.

EFTA, The effects of EFTA, EFTA Secretariat, Geneva, 1969.

EFTA, The trade effects of EFTA and the EEC 1959-67, EFTA Secretariat, Geneva, 1972.

El-Agraa, Measuring the impact of economic integration, University of Leeds Papers, 71, 1-18, July, 1979.

Erlenkotter, Donald, Economic integration and dynamic location planning, Swedish Journal of Economics, 8-18, 1972.

Glejser, H., Goossens, X., Vandeneede, M., Inter-industry versus intra-industry specialization in exports and imports, Journal of International Economy, 12 (3-4), 363-369, 1982.

Grubel, Herbert G., The theory of international capital movements - in: Black, J. and Dunning, J. H. (eds) - International capital movements, MacMillan, 1982.

Hallet, E. C., Economic convergence and divergence in the European community: A survey of evidence - in: Hodges-Wallace (eds) - Economic Divergence in the European Community, George Allen & Unwin - London, 16-31, 1982.

Harrod, Roy F., International Economics, Cambridge Economic Handbooks, VIII, Cambridge, 1933.

HMSO, Britain and the European Communities, HMSO, 1970.

Johnson, H. G., An economic theory of protectionism, tariff bargaining and the formation of customs union, Journal of Political Economy, 73, 256-283, 1965.

Johnson, H. G., Trade diverting customs unions: A comment, Economic Journal, 618-621, 1974.

Jones, A. J., El-Agraa, A.M., The theory of customs union, Philip Allan, Oxford, 1981.

Jones, L. P., The measurement of hirschmanian linkages, Quarterly Journal of Economics, II, 7-18, 1976.

Kaldor, Nicholas, The dynamic effects of the common market - in: New Statesman, pp 59-91, 1971.

Kaldor, Nicholas, The irrelevance of equilibrium economics, Economic Journal, 1237-1255, 1972.

Kaldor, Nicholas, The role of increasing returns, technical progress and cumulative causation in the theory of international trade and economic growth, Economie Appliquee, 34 (4), 593-617, 1981.

Keeble, D., Owens, P., Thompson, C., Centrality, periphery and EEC regional development, HMSO - Department of industry, London, 1982.

Kiljunen, Marja Liisa, Regional disparities and policy in the EEC - in: Seers et al. (ed) - Integration and unequal development, 199-221, 1980.

Koester, U., The redistributional effects of the common agricultural financial system, European Review of Agricultural Economy, 4 (4), 321-345, 1978.

Krause, L. B., U. S. imports, 1947-58, Econometrica, 221-238, April, 1962.

Krause, L. B., European Economic integration and the U. S., W. Brookings Institution, 1968.

Krauss, M. B., Recent developments in customs union theory: An interpretive survey, Journal of Economic Literature, 413-436, June, 1972.

Kravis, I. B. et al., ICP, phase III: International comparisons of real product and purchasing power, John Hopkins University Press, 1982.

Kreinin, M. E., Price versus Tariff elasticities in international trade - A suggested reconciliation, American Economic Review, 57 (4), 891-894, 1967.

Kreinin, M. E., The static effects of EEC enlargement on trade flows, Kyklos, 34 (1), 60-71, 1981.

Kuznets, S., Quantitative aspects of economic growth of nations: distribution of income by size, Econ. Develop. Cultural Change, II, 1-80, 1963.

Laffargue, Jean-Pierre, Les investissements americains dans les pays industrialises, Presses Universitaires de France, 1974.

Leibenstein, H., Allocative efficiency versus X-Efficiency, American Economic Review, June, 1966.

Lipsey, R. G., The theory of customs union: A general survey, Economic Journal, 60, 496-513, 1960.

Lubitz, R., Export Led growth in industrial economies, Kyklos, 26 (2), 307-319, 1973.

Lunn, John, Determinants of US direct investment in the EEC - further evidence, European Economic Review, 13, 93-101, 1983.

Lunn, John, Determinants of US direct investment in the EEC - revisited again, European Economic Review, 21, 391-393, 1983.

MacDougall, Role of public finance in European integration - Commission of the European Communities, Report of a study group, I - II, Commission of the European Communities, 1977.

Machlup, Fritz, A history of thought on economic integration, MacMillan, 1977.

Maillet, Pierre, The economy of the European Community - European Doc. 1-2, CEC., 1982.

Major, R. L., The Common market: production and trade, National Institute Economic Review, 21, 24-36, 1962.

Marjolin, Economic and monetary union 1980 - Report of the study group - Commission of the European Communities, I - II, Commission of the European Communities, Brussels, 1975.

Mayes, D. G., The effects of Economic Integration on Trade, Journal of Common Market Studies, 17(1), 1-25, 1978.

Mayes, D. G., EC trade effects and factor mobility - in: El-Agraa (ed) - Britain within the European Community - The way forward, 88-121, MacMillan, London, 1983.

McCombie, J. S. L., Economic growth, the Harrod foreign trade multiplier and the Hicks super-multiplier, Applied Economics, 17 (1), 55-72, February, 1985.

McCombie, J. S. L., Ridder, J. R., The Verdoorn's law controversy: some new empirical evidence using U.S. state data, Oxford Economic Papers, 36(2), 268-284, 1984.

Meade, James E., The theory of customs union, North Holland Publ. Co., Amsterdam, 1955.

Mendes, A. J. Marques, A balance of payments explanation of growth differences in western economies, 1985. mimeo,

Mendes, A. J. Marques, Economic Integration and Growth in Europe, Unpublished PhD Thesis submitted at the University of Kent at Canterbury, U.K., 1985a.

Miller, M. H., Estimates of the static balance of payments and welfare costs of UK entry into the common market, NIER, 57, 69-83, August, 1971.

Miller, M.H., Spencer, J.E., The Static Economic Effects of the UK joining the EEC: A General Equilibrium Approach, Review of Economic Studies, 44, 71-93, 1977.

Molle, Willem et al., Regional disparity and economic development in the European Community, Farnborough, Saxon house, 1980.

Moller, J. O. (ed), Member states and the community budget, Samfundsuidenskabeligt Forlag, 1982.

Mundell, R. A., International trade and factor mobility, American Economic Review, 47, 321-325, 1957.

Nugent, Jeffrey B., Economic integration in Central America - empirical investigations, John Hopkins University Press, 1974.

Ohlin, Bertil, Inter regional and international trade, Harvard University Press, 1933.

Paine, Suzanne, Replacement of the West European migrant labour system by investment in the European periphery, in: Seers et al. Underdeveloped Europe - Humanities Press Inc., 65-95, 1979.

Penketh, K., The cost to the U. K. acession to the EC, Journal of the Economic Studies, 7 (2), 99-108, 1982.

Pethith, Howard, European Integration and the terms of trade, Economic Journal, 346, 262-272, 1977.

Prewo, Wilfried, Integration and export perfomance in the European Economic Community, Weltwirtschafliches Archiv, 110 (1), 1-37, 1974.

Resnick, S. A., Truman, E.M., The distribution of west European trade under alternative tariff policies, Review of Economics and Statistics, 56 (1), 83-91, 1974.

Robson, Peter, Integration Development and Equity - Economic Integration in West Africa, George Allen and Unwin Ltd., 1983.

Robson, Peter, The economics of international integration, Studies in Economics: 17 - George Allen and Unwin 2nd ed., 1984.

Rollo, J. M. C., Warwick, K. S., The CAP and resource flows among EEC member states, Working paper, 27, Government Economic Service, London, 1979.

Scaperlanda, A., Balough, R.S., Determinants of US direct investment in the EEC - revisited, European Economic Review, 21, 381-390, 1983.

Schmitz, A., Bieri, J., EEC tariffs and US direct investment, European Economic Review, 3, 259-270, 1972.

Schmitz, A., Helmberger, P., Factor mobility and international trade: the case of complementarity, American Economic Review, 60 (4), 761-767, 1970.

Scitovsky, Tibor, Economic Theory and Western European Integration, George Allen and Unwin Ltd., London, 1958.

Secchi, Carlo, Impact on the less developed regions of the EEC, in: Seers et al. The second enlargement of EEC, 176-189, 1982.

Seers, D., Introduction - The periphery of Europe, in: Underdeveloped Europe - Humanities Press Inc., 1979.

Sellekaerts, W., How meaningful are empirical studies on trade creation and diversion, Weltwirtsch. Archiv, 519-551, 1973.

Shibata, H., The theory of economic unions: a comparative analysis of customs unions, free trade areas and tax unions, - in: C.S. Shoup (ed) - Fiscal Harmonization in Common Markets, Vol. I - Columbia University Press, 145-264, New York, 1967.

Steinherr, Alfred, Convergence and coordination of macroeconomic policies: some basic issues, European Economy, 20, 71-110, Commission of the European Communities, Brussels, 1984.

Stern, R. M. et al., Price elasticities in international trade: an annotated bibliography, MacMillan, London, 1976.

Theil, H., Economics and Information Theory, North-Holland Pub. Co., Amesterdam, 1967.

Theil, H. et al., The information approach to the prediction of interregional trade flows - Report 6507 of the Econometric Institute of the Netherlands School of Economics, 1965.

Thirlwall, A. P., The balance of payments constraint as an explanation of international growth rate differences, Banca Nazionale del Lavoro Quarterly review, 45-53, 1979.

Thirlwall, A. P., The Harrod trade multiplier and the importance of export-led growth, Pakistan Journal of Applied Economics, I (1), Summer 1982.

Thirlwall, A. P., Dixon, R.J., An exported-led growth model with a balance payments constraint - in: J. Bowers (ed) Inflation, development and integration: Essays in honour of A. J. Brown, 173-192, Leeds University Press, 1979.

Thirlwall, A. P., Hussain, M.N., The balance of payments constraint, capital flows and growth rate differences between developing countries, Oxford Economic Papers, 498-510, March, 1982.

Thursby, J., Thursby, M., How reliable are simple, single equation specifications of import demand?, Review, Review of Economic and Statistics, 56 (1), 120-128, 1984.

Tovias, A., Ex-post studies of the effects of economic integration on trade: Problems in measuring trade flow and welfare effects, Journal of European Integration, 5 (2), 159-167, 1982.

Truman, E. M., The European Economic Community: Trade Creation and Trade Diversion, Yale Economic Essays, 9(1), Spring, 1969.

Tsoukalis, Loukas, conomic divergence and enlargement, in: Hodges - Wallace (eds) - Economic Divergence in the European Community, George Allen & Unwin, London, 151-166, 1982.

Verdoorn, P. J., A customs union for western Europe - advantages and feasibility, World Politics, 482-500, July, 1954.

Verdoorn, P.J., Schwartz, A.N.R., two alternative estimates of the effects of EEC and EFTA on the pattern of trade, European Economic Review, 3, 291-335, 1972.

Verdoorn, P. J., Slochtern, F.J.M., Trade creation and trade diversion in the Common Market - in: Integration Europeene et realite economique, College d' Europe, Bruges, 1964.

Verdoorn, P.J., Van Bochove, V.A., Measuring integration effects: a survey, European Economic Review, 3, 337-349, 1972.

Viane, J.M., A customs union between Spain and the EEC: an attempt at, quantification of the long-term effects in a general equilibrium framework, quantification of the long-term effects in a general equilibrium framework, European Economic Review, 18(3), 345-368, 1982.

Viner, Jacob, The customs union issue, Carnegie Endowment for International Peace, New York, 1950.

Waelbroeck, J., Une nouvelle methode d'analyse des matrices d'echanges internationaux, Cahiers Economiques de Bruxelles, 21, 93-144, 1964.

Walters, A.A., A note on economies of scale, Review of Economics and Statistics, 45, 425-427, 1963.

Whitbread, Michael, Growth determinants in the regions of the EEC - an empirical study, Annals of Regional Science, 15(2), 13-26, 1981.

Whitby, M. (ed), The net cost and benefit of EEC membership - A workshop report, Centre for European Agricultural Studies, Wye, Kent, 1979.

Williamson, J., On estimating the income effects of British entry to the EEC, Surrey papers in economics, 5, 1971.

Williamson, J., Bottrill, A., The impact of customs union on trade and manufactures, Warwick Economic Research Papers, 13, 1971.

Winters, C.A., British imports of manufactures and the common market, Oxford Economic Papers, 36, 103-118, 1984.

INDEX